SURPLUS - 4
LIBRARY OF CONGRESS
DUPLICATE

ENERGY:
The New Era

ENERGY:
The New Era

S. David Freeman

Walker and Company · New York

First published in the United States of America
in 1974 by the Walker Publishing Company, Inc.

Published simultaneously in Canada by Fitzhenry
& Whiteside, Limited, Toronto.

ISBN: 0-8027-0460-3

Library of Congress Catalog Card Number: 74-77980

Printed in the United States of America.

Book designed by Stephen O. Saxe

10 9 8 7 6 5 4 3 2

Acknowledgments

Shortly after leaving government in August, 1971, I began work on this book under the sponsorship of The Twentieth Century Fund. Without its financial support, criticism, and assistance, this work would not have been undertaken or completed. During the winter of 1971, the give-and-take in my class at the University of Pittsburgh played a useful role in the formative stages of many chapters. Indeed, all my colleagues over the years have contributed unwittingly, for the book largely reflects what I have learned from them.

I am deeply grateful to Joy Dunkerley for her substantial contribution to the manuscript. She did much of the research and a great deal of analysis and filling in of many gaps. In the early months, Edwin Rothschild and later Philip Berardelli assisted in the research.

The manuscript was typed first by LaVerne Carlson, and then again by Billie Truesdell with speed, skill and encouragement to the author. My secretaries Annette Burke and Norma Dosky rendered invaluable assistance.

Marc Roberts was most helpful in reviewing an early draft as an economic consultant but bears no responsibility for the final product. And finally, I acknowledge with deep appreciation the editing of Richard J. Whalen whose work vastly improved the readability of this book.

To Marianne and our children—
Anita, Stan and Roger

Contents

PART I

INTRODUCTION

CHAPTER 1

The Making of a Crisis

October 17, 1973, was energy Pearl Harbor day. Instead of dropping bombs, a handful of oil-rich Arab nations shut off a few valves and sent shock waves through the closely linked high-energy industrial civilization in the United States, Western Europe and Japan. At first, few people in the U.S. realized that a full-fledged energy crisis was at hand. Oil tankers already at sea before the shut-off permitted normal deliveries for several more weeks. But by Christmas 1973, Americans knew the joyride was over.

On January 1, 1974, the oil producing nations dropped the other shoe, so to speak, by more than doubling the price of crude oil. To the energy-starved industrial nations the prospect of paying billions more coupled with continuing shortages of oil put the entire industrial world economy in jeopardy. And for the Third World the runaway fuel costs dealt a devastating blow to their budgets for development, which threatened to stunt their already slow growth.

America and the rest of the industrialized world had been drifting toward an energy crisis for a decade. The first warning to Americans of their heavy dependence on energy was the big blackout in the Northeast in November of 1965. There were other signals—power shortages almost every summer and shortages of fuel oil, natural gas and even gasoline in 1972—but these didn't affect the entire nation. The federal government continued to emulate Rip Van Winkle in its attitude toward energy. And the ordinary citizen who was still offered trading stamps to buy more gasoline and received sales pitches to "live better electrically" had every reason to scoff at the notion that the U.S. was running out of energy.

The Arab oil boycott, which cut off less than 5 percent of our total energy supply, simply telescoped the future for America. Shortages had already been growing more acute each year. America lived in a fool's paradise, believing it could continue a big car/glasshouse/energy-squandering industrial society with no thought for where energy came from, how much more was available, or how fragile our energy lifeline had become.

America's energy policy amounted to a blind act of faith that the oil companies and the utilities would indefinitely continue to deliver the goods. True, oil company ads warned us that "A nation that runs on oil can't afford to run short." But the remedy of higher prices for the companies and more polluted air for us simply confirmed the suspicion of many consumers that the shortage was a fake and a rip-off. Even after the fuel shortages became a reality in the winter of 1972-73, the federal government played down the seriousness of the problem. "Praying for mild weather" was the government energy policy, a risky policy at best, and a disaster for an administration whose credibility was at an all-time low.

It is difficult to fix responsibility for the grave error of failing to assure adequate fuel supplies for America. Government policy was to rely on the oil companies, and their policy was to put profits ahead of the public interest. After all, the oil companies are not public utilities and they recognize no obligation to build refineries, drill more extensively for oil and gas, or spend money on developing alternative sources of energy if they believe the price is too low or environmental-protection constraints are too stringent. Over the years, the possibility of an Arab boycott had been a good argument for maintaining the depletion allowance and quotas on imports, but it obviously was not a sufficient incentive to build up a costly oil stockpile in the United States in the absence of a government requirement. For years, Americans had lived with the unreal expectation that private companies would behave like the government while the federal government had slumbered on, lulled by soothing assurances from the industry as late as the mild "energy crunch" in the winter of 1972-73.

In the winter of 1974 the nation reaped a bitter harvest from this shameful neglect of the public business. The world moved toward an energy recession that might last several years. No one knew how high unemployment might go or how long it would take to find a way out. We did know these facts: Oil is the source of almost half of the energy in the United States and more than half in Western Europe and Japan; and the Middle East, by far the largest source of world oil supply, is of crucial importance to countries without a domestic supply. A sudden reduction in supply such as caused by the Arab boycott meant a sudden reduction in economic activity unless and until wasteful uses were cut back to make up the deficit.

Ironically, the fact that the U.S. wasted so much energy meant that with strong government leadership we could weather the crisis with a minimum of economic dislocation and unemployment. But a country such as Japan saw an oil boycott as a full-fledged disaster. Most of Japan's oil is used for industrial production. A cutoff meant industry cutbacks, unemployment and consumer panic. World trade linked the economics of the industrial nations. A decline of production in Japan and Western Europe intensified the economic slowdown in the United States.

America is lucky in the sense that the energy crisis cannot paralyze the nation. But it can cause hardship, anger, and even panic. We are faced with more than an energy crisis; it is a crisis and a test of our basic values. Can this nation keep its cool while the key elements of the affluent society—the big car and the big house—are "trimmed down" to the size of our energy supplies? If we can, the crisis could be the luckiest thing that ever happened to this country. It may be a painful experience, for we are witnessing the end of an era of cheap and abundant energy and all that this implied. Yet the need to concentrate on saving energy may cause us to face up to such problems as urban decay, suburban sprawl, and an enforced decline of mass transit. We may even find that the actions needed to balance our energy budget could remove much of the stress and tension that oppresses our lives.

The energy crisis could be a turning point in determining America's pattern of growth, a time for reassessment and a shift to a slower and more rational pace of activities. We could stop racing through our lives to see how we finished in the competition and develop lifestyles that require less gasoline and provide more time for enjoying our friends and neighbors as well as the beauty of nature. But in order to build this America and this way of life we must find out what caused the current crisis and how we can buy enough time to make a smooth transition to an energy-conserving Post-Industrial society.

The ordinary citizen is rightly bewildered, indeed angry. How could a government that spent billions of dollars to put a man on the moon allow its citizens here on earth to run short of fuel? And the oil companies with their many tax benefits and protection from foreign competition—what is their excuse? Is there still plenty of fuel in the ground? Has the oil industry been on a sit-down strike? Is the "crisis" a cruel hoax to charge us more and get rid of pollution controls? Who is really responsible?

The energy crisis is not a giant conspiracy concocted in a smoke-filled hotel room by politicians and oil company executives. True, the energy companies stand to profit handsomely from the shortage and they will take advantage of the crisis atmosphere to gain higher prices, eliminate pollution controls, and pursue corporate advantage where the public interest clashes with the industry's special interest. But there is no mystery about why we are in deep trouble. The record of industry dominance of government policy amid public indifference is an open book. The crisis results from a failure of private, corporate energy policies originating in Houston and Dallas and New York and rubber-stamped over the years by the Congress and a succession of Presidents. But the present scale of shortages is certainly not to the industry's liking. The industry knows that when consumers can't buy as much gasoline as they want, they demand action by the government, and such action could strike the industry a death-blow.

It was the depression of the 1930's that brought the government into the electric power business with the Tennessee Valley Authority and hydro-projects in the West. A depression caused by energy shortages could easily result in the federal government going into the oil business, especially since the government owns most of the economical fuel left in the ground. In fact the U.S. is one of the few countries without a government oil company.

The petroleum industry knows that it has "all its chips on the table" in the next few years. It has done well in the shadows for a long time because no one really cared about energy. But the kind of attention energy policy now receives means that people are watching how Congressmen vote and how the President acts on energy policy. The public interest flourishes in the spotlight.

The energy crisis arises from a more fundamental problem than the behavior of the energy companies. To fashion a policy for the future we must understand the basic realities and the clash of values that are at the root of the matter. The most important reality is that America has been devouring energy as though it were as plentiful as water in a rainstorm. The savings we were able to achieve during the winter of 1973-74 are ample proof of our lavish waste of energy, encouraged by industry promotion with government support. Cheap, abundant energy was taken for granted. For the past decade America increased its consumption of oil faster than it increased production. We filled the ever-growing gap by "eating off the shelf" until domestic capacity was fully put to use and then by increasing imports. At first the imports came from Western Hemisphere countries such as Canada and Venezuela, but in 1972 the oil necessary to sustain our rapid rate of growth began to come from the Middle East. Within less than two years, this U.S. dependence produced the crisis challenging the independence of American foreign policy.

America still has billions of barrels of oil and more billions of tons of coal in the ground. Why can't we simply switch back to domestic sources of fuel? Part of the reason is that we've already consumed much of the easily accessible oil; new sources are mostly off shore, where it takes longer to find and produce the oil. But a more fundamental constraint on the pace of production is a concern for protecting the environment. If the nation turned the fuel producers loose and let them charge as much as they pleased and drill where they pleased, we could have plenty of fuel in a few years. But at what cost? It would mean ravaging America to continue the joyride.

The U.S. is finally setting in motion laws to protect our beaches from oil spills, our hillsides from ruinous strip-mining, and our air from poisons. To turn back the environmental-protection clock is an unacceptable option. But therein lies our dilemma because we must accept the consequences in the form of reduced energy supply of our switch from "dirty" to "clean" fuels. Unfortunately America has invested very little time and effort in developing the technology to burn coal cleanly or to perfect improved sources of energy. And we failed to foresee that such research and development requires many years.

A major lesson of our present predicament is that energy shortages in a very fundamental way are a by-product of how and where people live and work—our lifestyles. Over the past few years America has continued patterns of growth in housing, transportation and industry that encouraged lavish consumption of energy while environmental policies outlawed dirty coal and slowed oil and gas production. Under the philosophy "more is better," we expected to go on doubling total energy consumption every 15 years. With a single-minded all-out fuel production program, we could fuel that kind of growth for a decade or two, but ultimately the clash of values must be faced. At stake are our beaches, the natural beauty of the countryside, and our right to breath decent air.

This clash of values can become very personal. How many people connect the big cars they drive alone to work with the pictures of oil spilled on the beaches? Or their air-conditioned houses with ruined mountainsides in Appalachia? And how many connect our lavish use of energy with the billions of people who cannot afford any of the comforts of a high-energy civilization? We cannot expect to remain energy gluttons, drawing rapidly increasing supplies from abroad, in a world where many nations are scrambling for the same limited supply.

Although the end of the Arab boycott brought some temporary relief, it would be a grave mistake to believe that imported oil can ever solve our problems. The lesson of the boycott was fundamental. The choice was not to give up our Cadillacs or give up on Israel. The Arab nations, with small populations and a desire to stretch out the lifespan of their only resource, are trimming their oil production levels to their own economic advantage. They are controlling production much the same as American companies once did in Texas.

The monopoly power of Middle Eastern oil-producing nations that now enables them—and in turn the multinational energy companies—to charge high prices for their oil accentuates yet another new reality in the fast-changing American energy situation. After being accustomed to low-priced energy for decades, the American consumer now confronts skyrocketing prices. This feeds the suspicion that the oil companies are getting rich off the energy crisis.

The money at issue is enormous. Each additional penny on the price of gasoline transfers a billion dollars a year from the consumers' pocketbooks to the oil companies' treasuries. Obviously clean energy costs more than dirty, and higher costs of labor and materials must be reflected in higher prices for energy. Citizens may be persuaded to accept price increases reflecting increased costs, but increases reflecting increased profits are hard to take.

Consumers naturally desire to buy energy at the lowest possible price. Yet higher prices are a powerful tool for cutting out the waste in energy consumption and encouraging greater production of new sources of supply. Obviously, higher energy prices will encourage better insulated houses, cars with improved mileage and more efficient use of energy in industry. Saving money by saving energy is likely to be a far more effective incentive for conservation than appeals to patriotism and civic pride.

An alternative is to raise prices, encouraging more efficient energy use, by imposing taxes on energy that could fund research and development of new sources and new technology. This approach has the advantage of insuring that government rather than industry captures any windfall profits. On the other hand, industry would not have the same incentive to increase supplies. Congress has not been eager to assume responsibility for raising energy prices. If it does so, the taxes should be progressive. Lower income groups could be badly hurt by high energy prices, no matter where the money goes.

The energy crisis we face is as deeply engrained as our lifestyles and the solutions mean reconciling changes in lifestyle with protecting the environment, maintaining our independent leadership role in the world and striking a fair balance between the prices consumers pay and the profits the companies make.

A crucial question that must be answered in laying out a realistic energy policy is the proper division of responsibility between government and industry. Ordinarily the marketplace is the desirable place to balance supply and demand because it is usually the most efficient. And there are still many people who believe that a purely market-oriented solution is the best way to overcome the energy crisis. The problem is that a free market economy pays more attention to short-term benefits than to the long-term social and economic costs that are at the heart of the energy crisis. The marketplace doesn't pay much attention to the future because investments that don't pay out for many years are seldom profitable to a private company and therefore aren't made. Investments such as controlling pollution that are most profitable to society don't bring any return to a company's shareholders and may cut its earnings. Indeed, on a discounted cash-flow basis the earth simply is not worth saving. By the same token, the research and development necessary to perfect new sources of energy is not an investment that any company or group of companies can undertake because the sums required are huge, the pay-off is decades away, and the benefits accrue to society rather than the investors. For several decades, government policy left environmental protection and energy research and development (except for atomic power) up to industry; the results were a polluted environment and complete neglect of new sources of energy supply.

Similarly the market places no greater value on oil from secure sources than on supply from the Arab nations that have boycotted the U.S. The oil companies have had no financial incentives to build up stockpiles in the U.S. that could have reduced our vulnerability, or to favor domestic production when overseas supplies were cheaper. The government's concern to uphold foreign policy objectives in the face of economic warfare obviously requires intervention in the marketplace, which is also a battlefield.

Concerning prices, there is serious doubt whether the fuel industry is sufficiently competitive to assure consumers of prices reflecting only a reasonable profit. The utilities are, of course, natural monopolies whose prices everyone agrees should be controlled by government in the public interest. A more sensible pricing system should be established reflecting the environmental costs of particular sources of energy and the added national-security value of domestic energy. Such a system would require government intervention in the form of pollution taxes and tariffs. It would, of course, take advantage of market forces to encourage energy conservation. But as we found out in the winter of 1973-74, balancing our short-term energy budget requires mandatory action by government. And in the future such major aspects of environmental policy as protecting the face of the earth and banning pollution harmful to human health will require mandatory government regulation.

A four-cornered relationship exists among low prices, adequate supplies, pollution control and foreign policy considerations, and the government must determine how these factors are to be balanced. If the price of energy is subsidized merely to keep its price low, then wasteful consumption patterns will continue and the problems of inadequate supply and pollution will be exacerbated. This in turn will mean increasing imports and consequently will cause international problems to loom larger. If the overriding policy objective is to minimize pollution in the U.S., we will need to start seriously conserving energy and paying the cost of the cleanup in higher energy prices. If the foreign policy problems arising from increasing imports are deemed of overriding importance, then we must take strong measures to reduce energy consumption, including higher prices and some environmental degradation, in order to obtain the needed supply from domestic sources. Thus, if we press down too hard on one policy objective the other three pop up. Judgments are necessary as to which goals have priority, and programs must be shaped to achieve our basic national policy goals.

There is still another fundamental reason why government cannot leave the energy problem to be solved by traditional market forces. The federal government owns most of the fuels that are likely to be economically recoverable for the foreseeable future. The federal government's gross failure to exercise its responsibility as the proprietor of these fuels, even to know what it owns, and to devise speedy means of bringing these fuels to market without ruining the earth is a major cause of our energy crisis.

In sum then, we face a crisis stemming from government inaction, from counter-productive policies, and from a mistaken faith that unguided private enterprise would be responsive to the public interest. It is now clear that adequate supplies of clean energy will require a massive federally financed research and development effort, and that balancing our long-term energy budget in a manner that implements anti-pollution safeguards and maintains U.S. freedom of action in foreign policy will be the business of government in the years ahead. Market forces surely must be used and competitive forces strengthened but the government must lead.

The central task of the government in developing a truly effective national energy policy will be to reconcile conflicting public goals. Energy conservation and developing new energy sources are basic solutions, but these will take time and cannot succeed without an overall policy.

Such a national energy policy will not be handed down by presidential edict or evolve from the marketplace. It will emerge from long debate and hard bargaining at all levels of American political life—local and state governments, the courts, both houses of Congress, the federal agencies and the White House. The stakes are high and the conflict in the political arena will be intense.

As America faces the new reality of scarcity, energy policy will be coordinated with national growth policy. A policy that requires energy conservation will influence many important aspects of modern life. It will affect the way we use the land, where and how houses are built, and the distance between homes and jobs; it will force the redesign of our transportation systems and the reappraisal of consumer and industrial products that are short-lived and wasteful. Energy policy will act as a mirror, reflecting the way we live and forcing us to imagine what the face of America ought to look like in the decades ahead. Once people begin to see how much energy they are wasting, they will find endless ways to economize and conserve, in the process solving many other problems afflicting the country. The stake in our search for an effective energy policy is nothing less than the way Americans will live and work for generations to come.

CHAPTER 2

The Age of Energy

As we suddenly discovered in the winter of 1973-74, our way of life is completely dependent on a reliable and adequate supply of energy. The fuel and electricity that is derived from oil, natural gas, coal, hydropower, and the atom have steadily taken over much of our daily work. Energy cooks, washes, dries, vacuums, heats, cools, and illuminates most American homes. It drives us back and forth to our jobs, takes us on vacation, and brings us most of our entertainment through TV, radio and movies. In the factories and commercial establishments where most of our energy is used, energy greatly assists the worker and enhances his productivity by powering machines and automated equipment. The fuels and electric power used in the United States provide every American with the energy equivalent of as many as two hundred full-time personal servants. Those people living in poverty consume less energy and possess relatively few such servants, but they too seek to join the high-consuming majority.

The Big Blackout in New York City and most of the Northeast late in the afternoon of November 11, 1965, first brought home to many Americans our near-total dependence on energy. Those who experienced the shock and fright of being trapped in elevators, in subways or in their offices, unable to get home, have not soon forgotten it. No one knew what caused the blackout. President Johnson heard the news on the radio at his Texas ranch, and his first reaction was to call Defense Secretary Robert MacNamara and be sure that our strategic defense systems were alert and operable. As the blackout affecting 30 million people persisted through the night, the President asked the Federal Power Commission, the agency that regulates the electric power industry, to investigate. The FPC was as stunned and unprepared as the average New Yorker, and the utilities affected were so unprepared that the lights even went out on the panels controlling the power plants themselves. The blackout of the Northeast could have been a major tragedy instead of an expensive inconvenience. Only a bright full moon prevented serious loss of life as airplanes landed without light. Even so, property damage from food spoiling in freezers, machines stopping without warning, computer time lost and a host of other causes was estimated in the millions of dollars. Although the statistics showed a measurable increase in the birth rate nine months later, most people did not find the blackout highly enjoyable.

The electric power companies since have been able to prevent recurrences of region-wide blackouts. But power shortages in one region or another have become regular events called brownouts. Voltage is reduced, which, within limits, has only minor effects such as reducing the quality of television pictures. If the brownout is severe, power is rationed and certain industrial customers may be cut off. If industrial customers cannot be cut off, demand is reduced by cutting off an entire area for an hour or two. The cut is rotated among areas so that inconvenience is spread more or less equitably. The brownout technique is used to reduce demand so that everyone's service doesn't blackout at once.

By 1972, the summer power shortage was accompanied by a gasoline shortage. And then these were accompanied by the winter fuel shortages, which first hit the Midwest in the winter of 1972. A year later, the nationwide energy crisis developed. Americans now receive year-round, inconvenient reminders of how much their comfort and essential services depend on energy.

When measured from a long-term perspective, modern society's current dependence on commercial energy has developed quite swiftly. As recently as 1850 people in the United States still obtained as much as two-thirds of their energy from human musclepower and draft animals. Wood was the main fuel, and almost all households burned it. For industry, the waterwheel furnished power directly and windmills supplied most of the mechanical horsepower. The amount of energy consumed was quite small. Indeed, all of the commercial energy consumed by mankind prior to 1900 probably would not equal this year's consumption alone.

The industrial revolution ushered in what may be called the Age of Energy. It got under way in this country in the 1860's, for the most part, and was marked by the increased use of the steam engine, which required great amounts of fuel, mainly coal. Lying dormant in the earth's crust, virtually untapped by mankind, were the coal, oil, natural gas and other sources of commercial energy needed to fuel the new technologies as fast as they were being developed. The chemical interactions and work of nature over a 100 million years were required to form fern fossils into these hydrocarbons. Thus, once used, they could not be replaced for at least another 100 million years. This is a sharp reminder that even a seemingly ample 300-year supply of fuel would last but a mere ten seconds in the long year of nature's complex and creative manufacturing process.

Coal was an important source of energy in the U.S. industrial revolution. Its major uses were for generating steam and in making iron and steel. Coal, of course, had been used in Europe for centuries and many small coal mines had existed in America since Colonial times. But as long as firewood was plentiful and mining machines and transportation were limited, coal was not used extensively. Until the early 19th century, even steam engines, river boats and railway locomotives were fueled by wood.

After 1865, coal came into its own. The remarkable and unparalleled growth in the production of coal from 20 million tons in 1860 to 500 million tons in 1910 powered the industrialization of the nation. During this transformation and redirection of a once agricultural economy, coal displaced firewood in most homes and industries. If we combine bituminous coal (which was used mainly in industry) and anthracite coal (which was preferred in households), coal had captured 90 percent of the energy market by 1910.

Coal ushered in a new era in American civilization and radically changed people's lifestyles. People moved into urban America to be near the factories that offered jobs. The coal-fired steam engine made possible large-scale operations that were much more economical than small plants, and big plants steadily replaced the small-scale, decentralized and subdivided manufacturing process. The steam engine not only brought economies of scale, it also became the epitome of progress.

This large-scale exploitation of America's coal seams marked a new turn in our civilization, the beginning of intensive exploitation of an asset of limited life in contrast to the renewable wood, wind, and water power of the past. As we drew on this irreplaceable resource capital, another turning point just as significant in human terms occurred, which Louis Mumford has described:

> Now, the sudden accession of capital in the form of these vast coal fields put mankind in a fever of exploitation: coal and iron were the pivots upon which the other functions of society revolved . . . The animus of mining infected the entire economic and social organism: this dominant mode of exploitation became the pattern for subordinate forms of industry . . . The psychological results of carboniferous capitalism—the lowered morals, the expectation of getting something for nothing, the disregard for a balanced mode of production and consumption, the habituation to wreckage and debris as part of the normal human environment—all these results were plainly mischievous.[1]

The oil industry also had its beginning just as the nation

started to industrialize. The world's first commercial oil well was drilled by E. L. Drake in 1859 near Titusville, Pennsylvania.[2] Interestingly enough, the first market for oil was as a medicine and the first mass market was as kerosene, a fuel for lamp light which replaced coal and whale oil. Oil also served as a lubricant essential to the smooth functioning of the new industrial machines. Within a few years, oil captured most of the U.S. lighting and lubricating markets and was being exported to Western Europe. During the Civil War, oil exports helped preserve the Union by replacing Southern cotton as an exchange for imports vital to the war effort.[3] For the next century, the U.S. oil industry dominated the world oil scene.

As Americans began to shift to the new ways of industrial life after the Civil War, John D. Rockefeller made his great fortune in oil. He consolidated his interests in a ruthless manner which left an indelible imprint of concentrated economic power on the entire oil business. His operation, even though quite efficient and most enterprising, stamped the oil industry as a strike-it-rich, public-be-damned monopoly.[4] But the early capitalist period also saw considerable growth and improvement in the technology of the industry. Refinery capacity initially was centered in Pittsburgh but it soon spread to major cities of the East Coast and expanded rapidly. Marketing and distribution systems benefitted from the introduction of such innovations as pipelines, first made of wood and then of iron and steel. The rail network, using rudimentary tank cars—again, initially made of wood—was extended into the oil fields.[5]

Even with all of the manipulations of the marketplace and the secret, anti-competitive dealings, the oil industry itself actually got off to a slow start as a major source of energy. In its first 40 years, the period 1860-1900, a total of only 1 billion barrels was produced—the equivalent of about 3 months of current production. Even so, by the turn of the century, the consensus was that we were "running out of oil." This attitude changed dramatically on January 10, 1901, when "big oil" first gushed from the earth at a spot in southeast Texas called Spindletop. The next year, Henry

Ford organized the company that ignited a gigantic new market for oil. The mass production of Model-T Fords and the first successful aircraft flight at Kitty Hawk by the Wright Brothers in 1903, launched America into the age of the motor vehicle and the airplane, two forms of convenient, comfortable and swift transportation that soon began to dominate our lives.

The remarkable period of technological innovation and economic expansion in the United States also witnessed the beginning of the electric power industry. Thanks to the ingenious inventions of Thomas Edison in the field of electricity, the first central, steam electric plant in the world began operating in New York City in 1882.[6] The kerosene lamp and the gaslight were thus quickly doomed to be replaced by the brighter, more convenient and apparently cleaner electric light bulb. Actually it was the electric motor much more than the light bulb that revealed the flexibility and versatility of electricity as a source of commercial energy. The electric motor spurred the growth of American industry by increasing efficiency and productivity in the country's manufacturing plants. Industry adopted electric motors with great speed, and by 1920 they accounted for more than one half of total installed horsepower in factories, compared with less than 5 percent in 1900.[7] These early uses of electricity were, of course, only the beginning of an endless array of appliances and conveniences and important new industrial uses.

Thus within a short period around the turn of the century, America witnessed the emergence of the major segments of what has become today's energy industry and the complex, fuel-hungry technology powered by these forms of energy. Once under way, the energy industry advanced quickly, and irresistibly. The world of the machine had arrived and with it the joys of unbounded material success, and the sorrows of lost values accompanied every clang, bang and roar.

Coal was still king in the production of energy in the early 20th century, but events began to take shape in peacetime and in war which led to the rapid conversion to oil. Winston Churchill, as First Lord of the [British] Admiralty just before World War I, was impressed with the vital role oil could play in developing a navy that was faster and able to stay at sea longer than coal-fired ships. As it happened, war broke out in Europe before his plans for exploration in the oil-rich Middle East could be implemented. The strategic importance of oil in World War I was established in other ways. By cutting off German oil supplies and war material, the Allied blockade hampered the newly mechanized German armies and severely handicapped needed war production.

After the war the U.S., impressed with the importance of oil, once again feared that its own oil sources had already been exhausted. In May, 1920, Senator James Phelan of California suggested the establishment of a federally owned company to search for oil abroad. In 1922, these fears were alleviated by major new discoveries in the U.S. and Venezuela. But the oil-rich Middle East was the real prize that interested all the industrialized nations and the international oil companies which dominated the market. In the post-World War I period the British and French became very active in trying to obtain concessions to explore for oil in the Middle East. In 1927 the vast oil resources in that part of the world began gushing forth in Iran. The largest U.S. companies, such as Standard of New Jersey, Mobil, and Texaco, as well as British and French interests gained a share of what became by far the major source of oil for the entire world.[8] The Middle East today contains 63 percent of the earth's known reserves.

Spurred by the new discoveries of oil and popularity of the automobile and other new oil consuming machines, the Age of Energy steadily shifted the focus of consumption from coal to petroleum. Motor vehicles and airplanes provided the major market for oil, but other markets also opened as railroads and homes switched to the cleaner and more efficient liquid fuel. This shift is still continuing as new discoveries in the Middle East and more recently in Northern Africa, the North Sea and Alaska attempt to keep pace with the remarkable growth in the consumption of oil.

Natural gas was another moving force in the transition from coal to petroleum during the past three decades. Natural gas is relatively new source of energy but the local gas companies that sell it date back to 1865, when gas manufactured from coal was first sold in significant volumes. Some companies are even older, but around 1865 improved technology gave birth to the modern gas industry by greatly reducing the cost of manufacturing gas from coal.[9] After World War II they simply switched to natural gas because it was much cheaper.

Natural gas was at first considered an unwanted byproduct in the search for oil. Thus, much of the gas was flared as the oil was produced, or if found in a separate well, it was shut in for lack of market. The early usage of gas was confined to the immediate vicinity of the fields and to a few nearby industries because of the lack of the technology to transport it very far. As pipeline technology improved, more and more residences and factories switched to natural gas, which was cleaner, easier to use and cheaper than coal.

An important breakthrough came in the 1930's: the development of large-diameter pipelines capable of withstanding the high pressures necessary to transport natural gas long distances. Pipelines could now economically carry gas many hundreds of miles.[10] A few of these early pipelines brought gas from the fields in the Southwest to the markets of Chicago, Minneapolis and Detroit.

World War II delayed the major expansion of the natural gas industry into other markets.[11] But in the post-war period, natural gas was delivered throughout the nation, and it rapidly replaced coal in homes and industry in the 1950's and 1960's. In fact, natural gas has supplied more than half the growth in total energy use in the U.S. since World War II. Today, fully one-half the nation's 63 million homes are heated by natural gas and most of the remainder by oil.

The most marked change in the American fuel picture over the past 20 years has been the decline of coal. In 1950 coal still provided almost 40 percent of this country's total energy consumption and was of major importance in both transport and industry, though already declining as a residential fuel. By 1960, coal's share had fallen to 23 percent and has since declined to 17 percent in 1972. Coal's main end-use now is as fuel for the electric generating industry. It has lost its place to oil and natural gas, which now account for 75 percent of the U.S.'s total energy supply.

Several factors lie behind these considerable shifts in fuel supplies. Changing technology, such as the dieselization of the railways which ended coal's role in the transport sector, have been responsible for some of the changes. Also, the fact that coal is dirty and inconvenient to store and handle has discouraged its use, particularly as a residential fuel. Recent environmental concerns have further contributed to coal's unacceptability.

In comparison to coal, oil and particularly natural gas, are clean fuels and are well adapted for use in some of the major energy end-uses—space heating and industrial processes. The phenomenal increase in the use of natural gas is also due to its relative cheapness, although there are great geographic variations in its price compared with other fuels. At the beginning of the 1950's, gas was one-third the price of coal, its cheapest competitor. Thereafter, the price of natural gas rose steadily until, by the end of the 1960's, it was only slightly cheaper than coal in many markets. But the environmental advantages of natural gas were sufficient to make it favored over coal—indeed, so favored that it cannot satisfy the demand.

The development of low cost transportation is an important element in the growth of our use of all forms of energy. Natural gas has, of course, benefitted from low-cost, large-diameter pipeline transport. Coal, though heavy and bulky, can still be shipped several hundred miles to market by train. Recently the unit-train, which carries coal non-stop from mine to power plant, has greatly enhanced coal's economic usefulness and caused it to grow in recent years as a power generating fuel. Similarly, the new super-tankers, capable of transporting 300,000 tons or more of oil, make it possible for Middle Eastern oil to move halfway around the world at relatively low cost. Extra-high-voltage electric power lines now enable us to locate generating plants outside of congested cities and near supplies of fuel and cooling water. But transportation, though it has created virtually a worldwide market, a market that keeps growing by leaps and bounds, cannot add to the finite supply of fossil fuels.

The commercial atom thus has entered the scene at a very opportune time. The power of the atom was, of course, first unleashed by the United States in the World War II Manhattan Project and then delivered against the Japanese cities of Hiroshima and Nagasaki. Out of a combination of guilt, pride, technological wizardry, and governmental foresight, we have since harnessed this power of the atom for the peaceful purpose of generating electricity.

The Atomic Energy Commission was established by the Congress in 1946 and given responsibility for both the development and regulation of atomic power. After a $3 billion, 25-year effort, the commercial atom came of practical and working age in the 1960's. But debate still rages over whether its poisonous by-products can be successfully controlled. Despite a good record to date there are still many unanswered questions concerning the future of atomic energy. Nevertheless, we are relying on the atom to supply most of the future growth in electric power supply, the only form in which we have thus far learned to harness atomic energy. And so the transition from wood to coal and then to petroleum as the main source of commercial energy now appears to lead to the atom, but this is probably not the last link in the chain. One day, we will finally turn directly to our planet's ultimate source of energy, the sun.

The energy industry in America developed swiftly because it was innovative and dynamic. Abundant sources of new fuels have been opened up even while supplies of traditional fuels were still plentiful. But the world has never experienced energy consumption on the present scale, and by the early 1970's the energy industry had become a victim of its own success. It tried to grow faster than its capacity in a time of environmental and political constraints at home and abroad. As a result, our high-energy civilization is in trouble.

PART II

PROBLEMS

CHAPTER 3

The Enormous Demands of a High Energy Civilization

Our increasing use of energy reflects the pattern of growth in the American economy. Total energy consumption rose in most years but fell in some. Through the 1950's energy consumption increased on the average by just over 2 percent annually—not much in excess of the 1.7 percent rise in the U.S. population.

Beginning in 1959, however, and continuing into the 1970's, energy consumption rose every year without exception, and the annual rate of increase doubled to 4 percent. Total energy consumption in 1972 amounted to 72 quadrillion btu's,[12] compared with 45 quadrillion in 1960 and 34 quadrillion in 1950. Numbers such as 72 quadrillion btu's are so enormous they are very difficult to visualize. Imagine a huge trailer carrying 22 tons of coal hitched behind every one of the 125 million cars, trucks and buses in the United States. This coal symbolizes our immense energy consumption in 1972.

Energy, of course, is not a homogeneous commodity. To understand this basic moving force of our high-energy civilization and economy[13], we must briefly examine the many types of energy, their characteristics, and the patterns of energy consumption.

Electricity is the best known form of energy but except for hydropower, electricity is not a basic source of energy. It is a convenient form into which an increasing share of petroleum, coal and atomic energy is converted. These fuels are "manufactured" into electricity at giant power stations resembling large industrial buildings. The fuels are converted to ash, gases and heat, and the heat content of the fuel is converted to steam which drives the machine that generates electricity. Transmission line systems deliver the electrical energy throughout the nation.

Electricity is now available to practically every home and farm in the nation.[14] On delivery to the consumer electricity is convenient, clean, and efficient. But these ultimate virtues cannot conceal the pollution of air, water and landscape involved in present methods of power production and transmission. In 1972 about 25 percent of the total raw fuel consumed in the United States was used for the production of electric power. Yet because of wasteful methods of production, electricity accounted for only about 8 percent of the total energy delivered to consumers.

Since World War II, the use of electricity has increased at an average rate of 7 percent a year in our homes, offices, and factories, a rate twice as fast as total energy consumption. This increase has arisen not so much from population growth—which, as noted earlier, has been under 2.0 percent per year—but primarily from mounting per-capita consumption. The relatively low price of electricity and its versatility no doubt account for the phenomenal increase in demand. To provide the capacity to meet this soaring demand, the electric power industry has become the nation's largest in terms of capital investment.

Electricity is the best known form of energy because of the many visible, identifiable and often-used products which consume electricity—light bulbs, refrigerators, TV sets. But three-fourths of all energy fuels are *not* converted into electricity. These are the refined products of crude oil and the natural gas and coal that are consumed directly in our homes, offices, automobiles, airplanes and factories. Although consumers are aware of the gasoline used to power their cars and the fuel oil used to heat their homes, few appreciate (or are even aware of) the vast quantities of oil thus consumed—the products of some 5 *billion* barrels of crude oil a year. This represents 42 percent of our total annual energy consumption.

Natural gas provides more energy than the entire electric power industry. But the housewife cooking with gas in New York City has no reason to reflect that her fuel has made a non-stop underground journey 1,500 miles from the bottom of the Gulf of Mexico off the coast of Louisiana. Similarly, the consumer rarely sees coal nowadays and thus has no reason to be aware of the mountains of coal burned annually in power generating plants and industry. In 1972 the United States consumed 520 million tons of coal—almost 60 percent of this amount in electric generating plants and the rest primarily in large industries. This represents 17 percent of our total annual energy consumption.

Although energy is all-pervasive in modern American life, the great bulk of the energy we consume goes into only a few end uses:[15]

- *Transportation* uses 25 percent of all energy, and probably 35 percent if we consider the energy required to make the cars, trucks and airplanes.
- *Space heating* of homes, offices and industrial establishments uses 18 percent of all energy.
- *Process steam, direct heat, and electricity to power machines* in industry use 36 percent of all energy.
- *Feedstocks* as raw materials for petrochemical plants to make a variety of items such as fertilizer, nylon hose, plastic bags use 5 percent of total energy.

This handful of major end uses accounts for more than 80 percent of total energy consumption.

Less than 20 percent of total energy consumption takes place in homes and apartments where people live. Americans use more energy traveling than in their homes, a very important fact in view of our grossly inefficient, almost chaotic national transportation system. More than half of the energy consumption in the United States used for transportation is consumed by the private automobile. About 14 percent of the nation's total energy consumption (and more than 30 percent of total petroleum supplies) goes to fuel privately owned automobiles.

The direct consumption of energy by ordinary citizens both in their homes and automobiles accounts for perhaps one-third (i.e., 24 quadrillion btu's in 1972) of total national energy consumption. How do we actually use most of the energy consumed in our households? A recent study[16] provides interesting insights. Those who habitually forget to turn off the lights will be comforted to learn that lighting is not a major consumer of household energy. Nor are the innumerable small gadgets such as blenders, toasters, carving knives, and coffee makers. Altogether these items may add up to 10 percent of residential energy used, or 2 percent of total energy consumed.

The major use of energy within the household is space heating which accounts for about 60 percent of total residential use. No other use is even close. Water heating (15 percent) is second and cooking and refrigeration third (12 percent combined). Except in very warm climates, air conditioning is not yet a major item; nationally only 13 percent of homes are air conditioned.

Residential consumption of energy has increased in every year since 1945. Consumption has grown more rapidly than population or the number of households, but in line with rising family incomes. The more affluent home has become more energy-intensive primarily because of increased per-capita consumption of electricity and fuels. Space heating requirements per person have increased as more households installed automatic heating plants and houses themselves became larger. Although space heating is important, its share in total household energy consumption did not grow between 1960 and 1968 as the use of electricity grew even faster. Air conditioners and clothes dryers accounted for much of this rapid growth.

About 15 percent of total energy consumption occurs in

office buildings, supermarkets and other commercial establishments. Air conditioning is already generally used in the commercial sector. Lighting is also a major consumer in this area. Even so, both of these uses combined are smaller than heating fuels, which account for more than half of commercial energy consumption.

The private automobile occupies a central place in the American lifestyle and economy. Consumption of gasoline has increased since the end of World War II by an average of more than 5 percent annually. The number of cars on the road has more than doubled since 1950, and the average American today drives about twice as many miles annually as he did in 1946, mainly because the population shift to the suburbs necessitates longer commuting journeys. Heavier cars also consume more gasoline per mile driven.

There is really little mystery why oil supplies are rapidly being exhausted: Americans are uniquely mobile. Private automobiles, buses, trucks, railroads and of course airplanes travel more miles and consume more fuel each year.

Nevertheless, as we previously observed, industry is the largest single consumer of energy in the American economy. Human labor provides less than 1 percent of the physical work performed in factories.[17] Huge amounts of energy are used in our factories, some 42 percent of total energy consumption.

Most industrial and commercial establishments, to be sure, are not heavy users of energy. Their energy use resembles residential consumption. They require fuel for heating and cooling and electricity for running machines. But energy amounts to only 2 or 3 percent of their costs. The great bulk of industrial use of energy is quite concentrated. Six categories of industry—primary metals, chemicals, petroleum refining, concrete-glass-clay-stone, paper, and food—account for more than 33 percent of the total consumption of energy in the nation,[18] which is half again as much as all the energy consumed in our homes. These six sectors account for 67 percent of total industrial energy consumption.

Total industrial consumption of energy rose steadily throughout the 1950's and 1960's. As would be expected, energy consumption in these years increased as industrial production expanded. Although the amount of fuel consumed by industry increased, it did not rise as rapidly as industrial production—that is, increases in production were achieved with a less than proportional increase in energy consumption. We received more output from each btu of fuel until the mid-1960s, when the trend reversed itself and the energy/GNP ratio began to decline.

Decisions by government and private business depend upon future expectations of energy demand. Thus despite the uncertainty there is no lack of energy consumption projections. One study[19] selected for review no fewer than 35 different projections, of which 15 were published during 1970 and 1971 alone. Many of these forecasts yield similar results at least through the 1980's. They reflect the rise in the rate of energy consumption that occurred in the late 1960's. They foresee the level of energy consumption rising from 72 quadrillion btu's in 1972 to 100 quadrillion btu's in 1980 and 125 quadrillion btu's in 1985. Beyond 1985 the estimates diverge. For the year 2000 the energy consumption forecasts range from 175 quadrillion btu's to 337 quadrillion btu's. The estimate of 1985 energy consumption of 125 quadrillion btu's implies that we will be able to find, deliver and consume in the next 15 years approximately twice as much energy as we found and consumed in the preceding fifteen years.

These estimates of future energy consumption could prove to be quite misleading. They do not really predict future demand as much as they merely extrapolate past trends in energy consumption. They ignore such important factors as changing trends in population growth and in the rate of growth and the composition of GNP. Moreover, they overlook the fact that energy prices are soaring rather than remaining stable as in the past.[20] Just as important, the forecasts ignore the possible impact of new government policies that might shift the U.S. away from promotion of greater energy use to conservation of energy.

Obviously population is one of the key elements in the size

of future energy demands. Since many of these projections were made, the U.S. Census Bureau has revised rates of population growth sharply downward. Birth and fertility rates are declining, and though zero population growth is still far off, rates of population growth are now significantly lower than those assumed in the projections. For example, most of the past energy projections assume a U.S. population of more than 300 million by the year 2000 yet one current estimate forecasts only 260 million.[21]

It is important to realize, however, that the downward revision in expected population growth will have only a moderate effect on energy consumption during the next two decades. People consume very little energy until they form new households. Thus the recent drop in the birth rate will not have an appreciable impact until about 1985. But beyond 1985 without compensating errors, the current projections are overstated by more than 10 percent on the basis of population alone.

Of course the major cause of increased energy consumption has not been more people but rather more consumption by each person. Continuing material affluence has meant using more energy in larger cars, larger homes, air conditioning and other conveniences and luxuries. The forecasters have correlated energy consumption with GNP, but here, too, history may not be the best guide for the future and mere projections may overstate future energy demand. First, the rates of growth in GNP used in many of the studies—some 4 percent per year—are higher than those achieved in the recent past and higher than many observers believe are likely to be achieved in the future. Second, the composition of the GNP may well be different from the past. If the economy expands less rapidly than anticipated and becomes more service-oriented, growth will be slower in the energy-intensive sectors and therefore energy consumption will be lower than projected.

The new attitude toward the size of families and even the formation of families could further upset forecasts based on extrapolation. The strong grass roots movement toward a greater simplicity of life and a concern for the environment carries with it important implications for slower energy-intensive economic growth. We have not yet even attempted to gauge the obvious impact of this trend on future projections of energy demand.

Another sharp break with the past is the developing shortage of fuels, in contrast to recent abundance. Most of the projections were based on an expectation of continuing abundant fuels supplies, but this may not be true in the future. Industry is bound to react to shortages and higher prices by finding ways to get by with less fuel. Widespread shortages of natural gas, heating oil, and gasoline will inevitably influence how much energy is consumed.

Rising prices of energy represent the sharpest break with the past and the one that is likely to play a crucial role in future patterns of energy consumption. Through many decades of low energy prices, America has developed lavishly wasteful patterns of consumption. The relationship of energy use to GNP established in periods of relatively low fuel prices will be a poor guide to the future when the price of energy is related to prices generally will be quite different. As later chapters indicate, there is every reason to believe that the price of basic fuels and electricity will continue to rise steeply for at least a decade.

The nation has little experience to gauge the effects of higher prices on energy consumption. Our past experience, and particularly the sharp rise in consumption occurring in the 1960's, occurred when prices of almost all sources of energy were either stable or rising more slowly than prices generally. In the recent past, energy prices, and the prices of many items which use energy, were becoming relatively cheaper. Given these conditions, consumers (including industry) reacted in the approved economic way—they bought more. The electric power industry has shown us in a dramatic way that promotional pricing of electricity does have an impact—lower prices have induced much greater use. Will much higher prices induce lower consumption? Are we so dependent on our cars and machines that a significant reduction even in the rate of growth cannot take place in a relatively brief period without major economic and social dislocations? Our experience of recent months suggests that savings can take place, but if they must be very large and on very short notice, the dislocations are quite painful.

Economists, recent experience and our own common sense all tell us that the reversal in the price trend of energy will have an impact on reducing the rate of growth of consumption—a factor omitted from the current array of projections. The tough question is how much of a reduction in overall energy demand will result from a given rise in price. The immediate response to modest price increases may be small. People are not likely to forego much air conditioning or heat because prices rise. And they are committed to existing houses, automobiles and machines. But over time, rising energy prices may encourage smaller cars, better insulated homes, and greater investments in efficiency by individuals and industry.

At the moment no studies have been completed that provide a solid answer as to the relationship between price and overall energy demand by 1985. A recent study suggests that electricity consumption would be sharply affected in future years by price increases.[22] But the weakness in such studies is their typically narrow focus: they examine only one form of energy and fail to reflect how much reduction in electricity use might increase oil or natural gas consumption.

The impact of price alone on overall energy consumption could be less than on particular forms of energy. All that can be said at the moment is that increased prices will have some impact, but until we gain more experience and complete more comprehensive studies we are traveling rather blindly into the future so far as our knowledge of the probable impact of higher prices is concerned.

Industrial use of energy is the area where higher prices and the fear of shortages are likely to have the greatest impact. The current state of the manufacturing art grows out of an era of relatively low energy prices. Industry has already begun to adjust to new conditions. As noted earlier the bulk of industrial energy is consumed in making metals, paper, cement, chemicals, and in a few other processes. These industries already have very large fuel bills and consequently must always seek ways to economize energy, particularly when prices rise or supplies tighten. Higher prices could lead to innovations which substitute new technologies that are much less energy-intensive. For example, new aluminum refining processes are under development that are claimed to reduce electricity consumption by as much as 30 percent. Such innovations would be entirely consistent with the country's dynamic industrial history.

These, then, are some of the reasons why the projections of U.S. energy consumption by the mid-1980's may be too high.

When we attempt to look beyond the next 10-15 years, forecasting becomes even more difficult. The speed of change in technology and patterns of living is quite rapid and produces unexpected results. It is no exaggeration to say that estimates for the year 2000 based on extrapolation of current trends are not worth the paper they are written on. By the year 2000 the structure of our economy may have changed fundamentally. By then we may be a service economy, with a proportionately smaller industrial sector than we have now. Our energy-intensive industries prospered in the past because energy has been cheap and abundant in the U.S. If energy becomes expensive here, businessmen may decide that it makes more sense to produce energy-intensive products in foreign countries where energy still remains comparatively cheap. The effect of many such decisions on the amount of energy consumed directly in the U.S. could be substantial.

Nevertheless, the projections serve a purpose: they compel us to confront the levels of energy supply necessary if future consumption continues past trends and patterns. In a sense they dramatize the problem and demonstrate that the nation must struggle to prove such forecasts incorrect in order to escape the serious consequences they imply. The future, we realize, lies in our hands. The policies the nation adopts in the next few years can shape the future energy-demand curves just as surely as the self-fulfilling prophecies of the energy companies encouraged past rapid growth. If the necessary slowing in energy consumption can be achieved through the working of the price mechanism, fine and good. If not, we cannot stand idly by. As a nation we must take policy steps to ensure that energy consumption fits within the supplies of clean energy likely to be available to us.

We use energy so wastefully in this country (see Chapter 10) that a significant reduction in energy consumption could be achieved rather painlessly, without sacrificing living standards. Much of our recent increase in energy consumption has been in uses which actually contribute only marginally to our increased enjoyment of life. As Barry Commoner points out, the increased consumption of energy doesn't necessarily mean that we live markedly better. Are we better off, say, because trucks haul freight instead of railroads? Or because we live at greater distances from our jobs and must drive more miles in bigger cars?

It would be wonderfully convenient if the price mechanism and market system would automatically bring about desired change. But will the system which brought us the non-returnable bottle and the glass curtain-wall office building be responsive enough to changed conditions to take these away again? Perhaps. But it would be foolish not to recognize the market's imperfections which may have to be removed before the price system can begin to help ease energy consumption (see Chapter 13).

So far we have examined energy consumption from a national point of view but it is also important to assess our future demands in a worldwide context. One reason is quite evident. The U.S. is obtaining a greater share of its supply from other nations. Our ability to import in the future will be greatly affected by the worldwide growth in demand for energy. Because we are all competing for and drawing on the same finite resources it is essential to examine trends and patterns of worldwide demand.

For many decades the U.S. was a leading exporter of energy fuels and still exports coal. By 1950, however, the net balance had shifted. By 1972 the U.S. imported almost 30 percent of its oil supply and would have imported more except for a system of quotas that limited oil imports. And imports of natural gas, although still only 4 percent in 1972, are expected to meet more than half of the growth in gas supply during the next decade.

The United States, having led the world into the Age of Energy, now finds itself increasingly dependent on other nations for sources of supply; though imports still account for only 15 percent of total energy consumption. But these same nations are also potential markets for our advanced technology in the energy field. We export very substantial quantities of enriched uranium for nuclear power plants abroad as well as a vast array of technology for power production and consumption.

Beyond adequacy of supply there are other reasons for taking cognizance of world energy demands. Obviously world market prices will be determined by the demands of other nations as well as the U.S. in relation to available supply. The implications for U.S. foreign policy of relying on energy supplies in unfriendly hands are enormous. And the economic, trade and balance of payments implications are also major considerations (see Chapter 6).

The worldwide growth in energy use has exceeded the U.S. rate of growth for at least 50 years. In fact the United States dominance of world energy consumption reached a peak in 1925 when the U.S. accounted for 48 percent of demand.[23] Since then, the share of the rest of the world in total consumption has materially increased. In 1970 the U.S. consumed 33 percent and the rest of the world 67 percent. But the unique extent of the American demand on world energy resources needs to be underscored to illustrate the problems these trends are creating. If all the people of the world had consumed in 1968 the same amount of energy consumed by the average American, world consumption in that year would have been 1,200 quadrillion btu's—almost six times higher than it actually was.

The Communist nations of Europe, led by the Soviet Union, have made particularly dramatic strides, growing from 7 percent of the world's total in 1925 to 26 percent in recent years. To a lesser extent, Latin American and African nations also have begun to increase their use of energy at a faster rate than ours, but their current usage is still quite small (4 percent and 2 percent of worldwide total consumption respectively).

Despite these rapid increases in consumption by the rest of the world, the U.S. is still the largest single consumer, currently accounting for about one-third of total world energy consumption. U.S. per-capita consumption is twice that of Russia, triple that of Western Europe, and 25 times greater than China.

The automobile, perhaps more than any other energy-consuming item, sets America apart from the rest of the world in terms of energy use. Of course, it is not simply the automobile but the entire American standard of living that accounts for our high use of energy. In no other country is there such emphasis on consumer convenience and personal comfort. A plentiful supply of energy at low prices has also encouraged the growth of many of our basic industries. U.S. agriculture, for example, leads the world primarily because it is the most highly mechanized, with energy supplying the driving force.

For the present at least, a close relationship still exists between a nation's GNP and its use of energy. Thus the U.S. with one-third of the world's GNP consumes slightly more than one-third of world energy supplies, Western Europe with 26 percent of GNP consumes 20 percent, and European Communist bloc countries with 19 percent of GNP consume 23 percent of world energy.

In Western Europe the consumer use of energy is following the American pattern but the average individual does not occupy as much housing space, does not own as many appliances, and drives a smaller car fewer passenger miles. Nor is there nearly as much demand for energy-intensive industrial products aimed at consumers such as plastic and aluminum foil.

The major use for petroleum in Europe is still as fuel and not yet as gasoline. Gasoline prices are much higher in Western Europe because gasoline is heavily taxed. Its retail price in Great Britain, France, Italy and other European nations has been close to $1.00 a gallon for many years which is one reason why Europeans use small cars to conserve fuel. This economizing has the further effect of curbing pollution as well.

The Soviet Union and Communist Eastern Europe form another major center of energy consumption in the world. Interestingly, the consumption of energy by these Communist nations now exceeds that of Western Europe even though in 1938 they consumed only 40 percent as much energy.

Energy consumption per person in the Soviet Union is now about 30 percent higher than in Western Europe. But this does not mean that the average Russian personally uses as much energy as his European contemporary. Private passenger cars, even small cars, are few in number in the USSR and account for only a tiny fraction of Russian energy consumption. The growth in energy consumption in Russia over the past half a century reflects the rapid pace of industrialization and urbanization. In a sense Russia's development parallels the American experience from 1860 to 1910. As a consequence, over the past four decades, Russia has expanded its share of the world energy market from about 2 percent to 18 percent. Even so per-capita consumption in Russia today is at approximately the same level as in the United States of 1905.

In Latin America, Africa, and Asia (except for Japan) the Age of Energy has not yet truly begun except in the larger cities and industrial complexes. In Africa and Asia particularly, life is still largely primitive and rural. The average person tills the soil with no machines to help him and dried cow dung may be his only source of heating fuel. In Asia, electrification is generally confined to the cities, although in China rural electrification has begun with the aim of installing one light bulb in each house and providing power for a radio. Most electricity is used for industrial purposes and even in urban areas it is used mainly for illuminating streets and public buildings. Home use of electricity is negligible.[24]

Asia contains 2 billion people, more than half the world's population, yet the region's total energy consumption barely equals U.S. consumption in 1925 and is only one-third of current U.S. consumption. Within Asia the disparity is equally great. Japan, a nation of 100 million people, uses as much energy as all of China with its 800 million people and consumes more energy than the 500 million inhabitants of India.

The contrast between the Soviet Union and China puts the future into rather dramatic perspective. In 1925 both nations consumed about the same volume of energy; today the Soviet Union is an industrial giant consuming three times as much energy as China and eight times more on a per-capita basis. Yet China is now making great progress toward improving the standard of living of its people and can be expected to consume more and more energy in the process. China possesses vast reserves of coal and it is developing a domestic oil industry.

If we assume that China in the year 2000 has a billion people and by then reaches the present Russian level of per-capita consumption, China's consumption of energy would increase ten-fold. China then would be burning the equivalent of 130 quadrillion btu's of energy per year, or twice the present United States consumption. If in addition we think of the billion people in India, Africa and Latin America and those who will be born despite the best efforts at population control, we see that the United States in the next three decades will face tremendous competition for the earth's limited material and fuel resources.

Estimates of future world growth in energy consumption, though allowing for some increase in per-capita use, do not foresee—at least in this century—worldwide consumption exceeding American levels of the 1920's. Nevertheless, the figures actually projected are staggering—350 quadrillion btu's in 1980 and 830 quadrillion btu's in the year 2000, four times current levels.

If the nations of the world actually consume these enormous volumes of energy from existing sources, major energy shortages and environmental disasters are in prospect. And if the rest of the world does not experience economic development and progress, we will live in the midst of worldwide poverty, discontent and danger. Against this background of enormously increasing demands for energy in all parts of the world, we now turn to the question of whether these vast demands can be met in the face of already serious and urgent environmental problems.

CHAPTER 4

Energy versus the Environment

The nation is at a critical cross-roads in dealing with the environmental problems created by the high use of energy. We are now painfully aware that our air and waterways are polluted and that more of the good earth is in danger of being destroyed by demands for more energy.

But the euphoric phase of discovering present and future threats to the environment is over. We are now embarked on the detailed, expensive, and exceedingly difficult task of implementing the environmental goals that our society has adopted. Undertaking the necessary clean-up effort brings us into conflict with the industrial and energy companies that are causing pollution and, even more fundamentally, it brings us into conflict with ourselves and with our habits and desires as a high-energy civilization. We are searching for a strategy that will enable us to switch energy systems from "dirty" to "clean" without sacrificing our way of life.

We now hear the argument from the energy companies that the environmentalists have helped cause an energy crisis. We are told that we must show "restraint" and "balance" in enforcing environmental laws—by which industry all too often means only delay in enforcing the laws. The case for delay in many instances seems plausible. There is a great temptation for our society to acquiesce and adapt to a deteriorating environment.

It is therefore necessary to look at the facts about the environmental problems which accompany high energy use.

First, we must face the fact that for the foreseeable future clean energy is a mirage. All sources of energy cause one kind of environmental problem or another. Yet society need not be seduced into abandoning environmental protection goals for this reason. The time has come to weigh the unhappy choices available and chart the most direct possible course toward cleaner energy. In developing this strategy we must be sure that we control the most serious problems, the real killers, and not dissipate our resources by tackling all problems indiscriminately.

Air pollution is already a direct threat to the health of mankind. It is caused almost exclusively by burning coal, oil, and to a lesser extent natural gas. Scientific evidence has linked existing levels of sulphur, carbon monoxide and very fine particulate matter to respiratory and heart ailments. The fuels burned in the automobile and truck in city streets and in power plants, refineries, and industry are the principal sources. Air pollution is thus a clear and present danger to man. We delay law enforcement to cut down air pollution at the expense of human health and possibly human life.

Atomic energy is the only existing source of energy that does not pollute the air (except for the very limited hydropower). But it poses a quite different kind of danger to man —the danger of radiation poisoning of his genes which is passed from generation to generation. Atomic power plants have a good safety record but there is the haunting fear of an accident that would release large amounts of the radioactivity contained in each plant and expose large numbers of people to excessive radiation. There are also the fundamental problems of where to store the "atomic trash" that remains a deadly poison for 25,000 years, and to make sure that weapons-grade material is not "diverted" into unfriendly or irresponsible hands.

Energy production poses a different set of environmental problems. Extracting coal from the earth can destroy the green cover over the land and the lifestyle of people in the despoiled region as well. And pumping oil out of the bottom of the sea and transporting it throughout the world raises a growing threat of oil spills that will pollute oceans and beaches. Much of the damage caused by energy production is local in nature. Strip mining disturbs the environment of the mountaineer in eastern Kentucky but not the consumer in Baltimore or Atlanta who uses this coal power when keeping cool or watching TV. And the people driving their cars in New York or Pittsburgh are not likely to worry about the life span of marine organisms in the Atlantic poisoned by oil spilled on the journey to their gas tanks. But we can no longer expect people in the coal country or along the sea coasts cheerfully to accept the burden of ravaged land or oily beaches so that the rest of us may continue to be energy gluttons. And yet for the nation as a whole the "skunk works" must be placed somewhere.

The critical question is how to shift from dirty to clean energy as rapidly as possible. Let us consider the problem of air pollution first and the heart of it—our dependence on fossil fuels.

Air pollution is the most damaging of current environmental threats to the public health. The Environmental Protection Agency estimates the annual cost of air pollution to health, vegetation and property values at more than $16 billion annually—$320 for each family of four in the U.S.[25] The fossil fuels have a common fault: When they are burned, air pollution results.

But there is considerable difference among these fuels as pollutants. Natural gas is the cleanest, coal the dirtiest. Oil is cleaner than most coal, but gasoline is the most significant offender in terms of volume of consumption: 100 billion gallons are burned on city streets and highways each year.

The pollutants currently discharged are often invisible. Air pollution control has generally eliminated the particles that can be seen in the form of black smoke. The air is now contaminated by gases in the form of oxides of sulphur, carbon, and nitrogen as well as very fine particulate matter of submicron size which are invisible to the naked eye. These pollutants combine in the air and damage the vegetation and materials they contact and the people who breathe the air.

The fossil fuels are not merely a part of the air pollution problem. Except for the burning of leaves or trash, and the wearing away of rubber tires, they are *solely* responsible for the entire problem. Unlike water pollution and land use problems where energy consumption is only one factor, air pollution is almost exclusively caused by discharges from fossil-fueled power stations, energy-intensive industries and automobiles, trucks and buses.[26]

The medical experts have linked air pollution to bronchitis, emphysema, asthma and lung cancer. It is not surprising that pollutants capable of disintegrating stone, corroding metal and dissolving nylon stockings also can damage delicate pink lung tissue.[27] The long term effects of air pollution on people that are now healthy are uncertain but worrisome. We know that the aged and those already suffering from lung or heart diseases are especially vulnerable.[28] The deaths that occurred during air pollution episodes in Donora, Pennsylvania (20 deaths in 1948), London (1600 deaths in 1952), and New York City (165 deaths in 1966) have proved that air pollution can be a killer.[29] Lave and Seckin in their study on the relationship between air pollution and mortality found that:

> A 50 percent decrease in air pollution is estimated to lower the mortality rate by 4.5 percent. Another way of translating the results is to note that if we assume that the reduction in air pollution would have the same effect on morbidity as on mortality (which is certainly a very conservative estimate), we could reduce the economic cost of morbidity and mortality by just over $7 billion per year.

The consumption of fossil fuels, especially in automobiles and trucks, is the chief source of the air pollution problem in cities in the United States and around the world. The automobile is the major threat in most cities but no single auto-related pollutant is alone responsible.[30] The air pollution problem seems to result from synergistic effects, whereby the combination of several pollutants causes more damage than would be expected from the sum of each individual pollutant. In addition to the discharges from burning fuels, the wearing away of rubber tires leaves an estimated 700,000 tons of material annually on U.S. city streets, about 10 percent of which is blown into the air we breathe. One ingredient of these fine rubber particles—carbon black—is known to be carcinogenic in certain circumstances.

Attempts to eliminate automotive air pollution have thus far centered on adding expensive gadgets to the combustion system to obtain cleaner burning. Unfortunately little or nothing has been done to achieve a more basic remedy. The nation has yet to face the reality that 100 billion gallons of gasoline probably cannot be burned within heavily populated areas without polluting the air and harming the people who breathe it. If only a small part of the effort that goes into attempting to clean up the internal combustion engine could instead be spent on a battery for an electric car or other alternatives, the air might become cleaner sooner. To be sure, electricity or other fuel sources would cause some pollution at the central generating station but it can be better controlled in a large plant remote from population centers. Moreover, the power could be provided from sources such as atomic, solar, or geothermal energy, which do not pollute the air.

Large coal- and oil-burning power plants are the other major source of air pollution. Technology is in commercial use to remove the sulphur from oil but not from coal. But both oil and coal plants emit the small particulate matter noted above.

> A 3,000 MW electrical station burning coal with 3-1/2 percent sulfur would emit 500,000 tons of sulphur oxide per year, 100,000 tons of oxides of nitrogen, and even with 99 percent efficient precipitation it will emit about 7,500 tons of fine particulate matter.[31]

Most coal-burning power stations each are throwing thousands of tons of harmful pollutants into the surrounding air each year. Unless remedial steps are taken, air quality standards simply will not be met.[32]

Sulphur has been identified as the chief poison emitted by power plants. Indeed the 25 million tons of sulphur discharged into the nation's air each year by such plants represent three-quarters of the total sulphur pollution. In addition, power plants emit about half of the oxides of nitrogen, about 10 million tons per year, and more than 7 million tons of particulate matter and unknown quantities of mercury and other harmful trace elements found in coal. Electrostatic precipitators can remove all of the visible particulate matter, but no control technology is being used to prevent invisible gases and tiny particles or trace elements from escaping into the air. Research and development programs are making slow and uncertain progress. For the foreseeable future, fossil-fueled power plants will continue to pollute the air. Even if low-sulphur fuels are used, the millions of tons of coal required each year at a large power plant overwhelm the air for miles around with excessive amounts of oxides of sulphur and nitrogen as well as fine particles which eventually lodge in the deep recesses of the individual's lungs.

The nation is committed to a strong program to assure clean air. And recent data indicate that in major U.S. cities the levels of sulphur oxides and smoke have been reduced. Regulations that require lower-sulphur-content fuels and the installation of control equipment for larger particles have been effective.[33]

Yet we cannot be certain how much progress the nation is making in protecting public health. While the levels of some pollutants are declining, we learn that the standards imposed by law may not prove strict enough. Ironically, the western low-sulphur coal now coming into use to clean up the air may have just the opposite effect. This coal is generally higher in ash content and lower in heating value, thus requiring the burning of more tons per kilowatt hour. The effect is to greatly increase the volume of fine particulate matter thrown into the air. Other damaging pollutants also are not being controlled. For example, recent studies by EPA scientists suggest that suspended sulphate particles are a more important health hazard than is generally recognized. To protect life, air quality standards may have to become even more stringent.

In 1972, the National Academy of Sciences warned the nation that breathing the air in urban America could damage one's health. The Academy pointed to the dangers of polycyclic organic matter (POM), which is formed when fossil fuels are inefficiently burned. POM is created largely by the internal combustion engine in automobiles and by coal-fired furnaces, burning of coal refuse banks, and coke production. As the Academy report said:

> There is clear evidence that airborne POM found in occupational settings—especially in relation to the products of burning, refining, and distilling of fossil fuels—is responsible for specific adverse biologic effects in man. The effects include cancer of the skin and lungs, non-allergic contact dermatitis, photosensitization reaction, hyperpigmentation of the skin, folliculitis, and acne.
>
> It appears that the incidence of lung cancer among urban dwellers is twice that of those living in rural areas; and within urban communities, the incidence is even greater where fossil-fuel products from industrial usage are highly concentrated in the air.[34]

Many concerned experts believe that these fine particles are the prime causes of pollution-related illness. Even the Clean Air Standards established by law for 1975 will not control the growing volume of these invisible killers in the air.

In addition to all the known and suspected pollutants caused by burning fossil fuels, there is an imponderable. Fossil fuels are mainly carbon and when burned they release carbon dioxide into the atmosphere. According to Dr. Lester Machta of the National Oceanic and Atmospheric Administration, continued and prolonged use of fossil fuels at projected levels will cause a 20 percent increase in the amount of carbon dioxide in the earth's atmosphere by the year 2000.[35] Carbon dioxide is not a pollutant as such but such increased concentration in the atmosphere nevertheless poses a threat to mankind. Carbon dioxide acts as a "one-way mirror," permitting the sun's rays to reach the earth's surface but acting as a barrier so that heat cannot escape. It could thus achieve a greenhouse effect.[36] If this were to happen and the temperature of the earth were to increase even 2° or 3° F. some scientists believe that it could produce substantial melting of the polar ice caps. Eventually, such melting could flood coastal regions.[37] Observations to date indicate that the carbon dioxide content of the air has been steadily increasing for more than a decade at about 0.2 percent per year[38] and it is believed this increase is due to the combustion of fossil fuels.

There is no immediate danger that carbon dioxide levels will affect temperatures on earth. Indeed, it is only a theory and by no means a certainty that climate will ever be affected by increases in carbon dioxide concentration. Even so, the volumes of fossil fuels expected to be consumed in the world in the next decade are vast. If these projections are coupled with an extrapolation of current trends one can postulate that carbon dioxide levels could, by the end of the century, reach heights that would translate this theory into a harsh reality.

Some scientists are much more relaxed about the prospect. They believe that even if all the world's fossil fuels were burned and the earth's carbon dioxide level doubled, then the 2° rise in the planet's temperature might do no more than complicate our lives.[39]

They suggest the higher temperature on earth would cause more water to evaporate from the world's oceans, creating a heavier cloud layer, which in turn would filter the sun's rays and then cause a lowering of temperatures. These scientists believe the net effect would be to achieve an equilibrium, though there could be extensive flooding before an equilibrium was reached.[40] Other scientists are concerned about an opposite trend, a cooling of the earth as particulate matter builds up in the air to shield the sun. And in fact there has been a slight cooling in recent years. If it continued, a lowering of temperatures of only a few degrees could cause another ice age over most of the lands where people now live.

All the theories about the effects of carbon dioxide and

particulate matter build-up are speculative. The fact is that we simply do not know what will happen. We can only be certain that man is tampering ignorantly and perhaps dangerously with this planet's environment in a very fundamental way. And if we find that excessive fuel consumption is causing threatening changes in climate, the lead time for reducing fuel consumption to ward off the threat will be quite short. The effects on a high-energy civilization without alternatives could be disastrous.

It is also important to understand that when energy is finally consumed in one form or another, it doesn't simply disappear. The energy becomes heat. And while the level of heat in our cities has not reached intolerable limits, at least for those who live inside air-conditioned structures, it is already affecting the urban climate. And as the air inside more homes is cooled, the waste heat from the air conditioners raises the temperature outdoors. If present patterns of concentration of population and greater per-capita consumption of energy both persist into the future, some informed observers fear that a city could become a "heat island" that would so alter the climate as to be unlivable.[41] At the moment, in the Los Angeles basin man-made heat is already equivalent to 5 percent of solar energy absorbed at the ground. By the year 2000 it could reach 18 percent.[42] Thus, even if a completely "clean" source of energy that is self-renewing were to be developed there may be a definite limit on the level of consumption which the environment can tolerate in a given locality.

The fossil fuels combine in polluting the air but each one presents environmental problems of its own in the process of being extracted from the earth, converted to a usable form, and delivered to the consumer. Now we shall examine some of the major sources of environmental danger.

OIL

Oil spills are as old as the petroleum industry, yet land-based production has been conducted in a manner that has largely prevented physical waste and minimized pollution.

True, the wells themselves are not a joy to behold, except perhaps to the owners. Salt water produced from wells is a problem, especially as ground water supplies become more scarce. And from time to time wells do blow out of control. But the number of blow-outs is small—only 100 in the last ten years and they were mostly gas wells.[43] By and large, land-based exploration and development of oil and natural gas have not caused serious environmental problems.

Drilling for oil in several hundred feet of water is quite another matter, however. Oil is spilled when wells flow out of control and when there are breaks in the pipelines that take the oil on shore. On the basis of numbers, the industry's record of drilling safely in water-covered areas is good. More than 14,000 offshore wells had been drilled through 1970 and only 25 had blown out of control.[44] But a considerable amount of oil has been spilled from those blow outs, other accidents, and the negligence of oil companies. The Santa Barbara spill of some 50,000 barrels, though the most noteworthy in recent history, is not the largest. Pipeline accidents also contribute to the pollution of our coastal waters: a single accident in the Gulf of Mexico in 1967 accounted for 167,000 barrels. More than 3,000 barrels of oil are discharged into the Gulf of Mexico each year in the waste water from off-shore operations.[45] And intensive exploitation of the off-shore waters is only beginning. Thus far we have exploited perhaps not more than 10 percent of the oil and gas beneath the waters off the United States shores.[46]

In addition to oil damage to beaches we are now learning that crude oil, and especially its most toxic fractions, is soluble in sea water. Most of the oil spilled cannot be recovered, and while our knowledge is far from conclusive, there are indications that hydrocarbons from oil spills may enter the marine food chain and become a hazard to marine life and perhaps even to man himself.[47]

The prospect that crude oil may pass through the food chain in much the same way as other persistent chemicals such as DDT is a matter of genuine concern in view of the growing volume of oil spilled in the oceans. And especially because crude oil contains potential carcinogens.[48]

Ocean-going tankers are the other large source of oil pollution in the oceans. Tankers transport about half of the world's oil production. In 1970 more than 8 billion barrels of oil were transported across the oceans, and the total increases by 5 percent each year.

Tankers spill oil in various ways. They collide and break up in storms. And they spill oil routinely by using sea water for ballast and then discharging oily water. The major spills such as the break-up of the tanker *Torrey Canyon* off the coast of England in 1968 attract widespread public attention. But the daily spills from routine operations and accidental discharge by ocean-going ships is probably the largest source of oil pollution of our oceans.[49] It is estimated that some 22 million barrels of oil are spilled into the oceans of the world each year merely from routine cleaning of bilges.[50]

Waste oil poured down drains or spilled outdoors is another major source of water pollution. The oil eventually washes into rivers and on to the ocean, the ultimate trash can for our throwaway society.

The oceans are vast and one might wish to believe that the oil spilled in them just disappears. But one citizen of the world who crossed the ocean not long ago on a papyrus raft, close enough to the water's surface to see what is happening, tells us that the oil is not disappearing. Sailing from the coast of Africa to North America in 1970, Thor Heyerdahl found oil clots on the ocean's surface on 43 days of his 57-day voyage. Heyerdahl's findings are confirmed by U.S. Government scientists who cruised the ocean on three research ships in the summer of 1971. They found massive globs of oil and clots of plastic in the Atlantic Ocean from Cape Cod to the Caribbean Sea. The pollution covered 700,000 square miles, an area more than 2-1/2 times as large as Texas. Better than half of the samples of young fish collected from surface water were oil-contaminated.

The question then is what can be done to prevent this oil from spilling. Radar and other navigation aids should minimize tanker collisions. And there are cleaning procedures whereby oil can be separated from ballast water before a ship discharges it into the ocean. These systems supposedly are in effect on 80 percent of the world's tanker fleet.[51] But accidents do happen and policing the tankers at sea is not an easy task. The large super-tankers, which are less maneuverable than smaller tankers, pose threats of super-spills.[52]

Oil spills in the ocean are a serious problem whether oil is imported or produced by drilling offshore.[53] Despite the best efforts of all concerned the increased volumes of oil produced offshore and the larger quantities transported by tankers are resulting in more, not less, oil spilling into the oceans. And it seems reasonable to assume that the more oil we produce and consume the more we will pollute the oceans and the coastlines they touch.

The oil spill problem in the United States could be minimized if the sources of oil on the North American continent were linked to refineries in a pipeline system or other overland means of transport. In the past, underground pipelines have proven to be a reliable and inexpensive means of transporting oil. About 150,000 miles of oil pipelines are buried beneath the U.S. surface[54] and spills have been negligible. But Gulf Coast crude oil is now transported to the East Coast by tanker, not by pipeline. And there is little reason to believe that land-based oil sources in the United States will be greatly enlarged, except for oil shale and synthetics, which will be discussed later.

With respect to Alaskan oil, if it were delivered overland to the lower 48 states, oil spills could be avoided in the coastal waters. The decision to build a pipeline across Alaska and transship the Alaskan oil to the West Coast by tanker, passes up a major opportunity to reduce the danger of oil spills in Pacific waters and along the beaches of the west coast of Canada and the United States. The controversial pipeline to be built across Alaska also demonstrates that pipelines are by no means certain to avoid creating environmental problems.

Pipeline systems may be out of sight but they still require a clear right-of-way as much as 100 feet wide. The land surface is excavated and often permanently scarred when the pipeline is constructed. The oil pipeline across Alaska is a vivid and rather extreme example of the environmental problems that an overland line can pose. The pipeline will not only disturb the fragile tundra and wildlife in an otherwise unspoiled area, it will also be endangered by earthquakes which could rupture the line and cause a massive spill. The Alaskan pipeline, and the proposed alternative through Canada as well, highlight a fundamental truth: No matter how oil supplies are obtained, there is inevitably an environmental price to be paid.

NATURAL GAS

Natural gas production has not aroused much environmental concern because gas doesn't spill. Nevertheless, it is important to remember that a good part of the remaining natural gas supplies in the United States are to be found in offshore areas. In offshore drilling, oil and gas are usually an inseparable package. Therefore, although some areas are more gas-prone than others, we face a dilemma: Approval of natural gas production automatically means approval of offshore drilling for oil, with its greater environmental risks. mental risks.

The transportation of natural gas poses a different kind of problem not present in an oil pipeline—the risk of explosions. Natural gas is transported under intense pressure of as much as a thousand pounds per square inch. A break or even a sizable leak in a high-pressure natural gas pipeline can turn it into a gigantic blow torch that destroys all human life and property in the surrounding area. Fortunately the safety record of the natural gas industry to date has been rather good. Still, there have been accidents that resulted in the loss of human life.[55] There is inherent danger in a system which contains more than 800,000 miles of pipe, much of which traverses densely populated areas. And every major city has a natural gas distribution system beneath the surface in which gas is under pressure of some 200 pounds per square inch.

The primary danger lies in the fact that much of the pipe has been in place for decades; the distribution systems under city streets are the oldest of all. Moreover, some of the high-pressure pipelines were built under less stringent standards than those in force today and sections may be corroding. Federal legislation was enacted in 1967 to provide strict safety standards, but enforcement has been lackadaisical. Thus far the question of natural gas pipeline safety has escaped widespread public attention because there have been few accidents. The future record could be gravely different unless the government requires the natural gas industry to replace older sections of their pipeline systems.

COAL

The extraction of coal from the earth presents a Hobson's choice. Mining underground endangers the miner but tends to preserve the surface—although subsidence is still a major problem. Strip mining the coal above ground tends to be safer for the miner but destroys the countryside. Because coal represents this nation's largest existing source of energy, these problems of coal mine safety and the misuse of land are central concerns for energy policy.

The strip-mining of coal is visibly destructive.[56] Giant earthmoving machines, as wide as an eight lane highway and standing 10 stories high, gouge the earth at the rate of 220 cubic yards a scoop.[57] These machines rip off the vegetation and topsoil covering the coal and deposit the debris in spoil banks. The exposed coal is extracted and shipped to market. The results are:

- Barren wastelands that are already twice the size of the state of Delaware.
- Water pollution from acids and corrosive compounds washed away from the exposed area into nearby streams that poison fish and wildlife.
- Floods caused by "spoil banks" eroding and clogging streams, thus reducing their carrying capacity.
- Personal and property injury to homeowners located below from slides caused by strip mining operations.

Strip mining is a fast growing source of coal supply, now accounting for about half of the total annual production. Most of the stripping so far has occurred in Appalachia, in the states of West Virginia, Kentucky, Pennsylvania, and Ohio. But strip mining has now spread to the Western states of Wyoming, Montana, New Mexico, and North Dakota, where 77 percent of the nation's strippable coal reserves are located, much of it low in sulphur content. Nearly one million acres of public and Indian coal land containing upwards of 10 billion tons of coal are already leased.[58] Western coal seams are thick and the ground cover is relatively thin. And so the coal rush is on.

"Reclamation" is the one-word answer to the problems posed by strip mining. But as an environmentalist points out, " 'Reclamation' is without question the slipperiest word in the lexicon of strip mining."[59] Too often in the past reclamation has meant simply smoothing the overturned earth or planting a few seeds and hoping something would take root and grow. The failure to reclaim strip-mined land is an ugly fact beyond dispute.

The real issue is whether it is possible to halt the destruction of the countryside without forbidding the strip mining which causes it. Industry spokesmen claim that strip-mined land is in fact now being "reclaimed to productive uses."[60] In theory the land can be satisfactorily reclaimed by preserving the top soil intact, replacing it promptly as each area is mined, planting new vegetation on the land and nursing the new growth until it really takes hold. The theory, however, breaks down when the land is mountainous and the strip mining cuts a series of high walls and trenches around the mountainside. Satisfactory reclamation is very expensive and virtually impossible on steep slopes. Erosion, acidification of streams, and landslides seem inevitable. The theory may also break down in the arid West where ground cover is thin and true reclamation may be impossible. Lack of rainfall in many parts of the West makes it extremely difficult to reclaim the land after strip mining. No one is sure but the federal government isn't even trying to find out. In 1969 the Interior Department adopted regulations whose purpose was to prevent a recurrence in the West of the sad fate that strip mining inflicted on Appalachia. But the Department has failed to enforce its own regulations.[61]

There is little doubt that in level and rolling countryside

reclamation is possible if there is good soil and adequate rainfall. Examples of successful reclamation can be seen in parts of Pennsylvania and in Germany.[62] Despite laws in many states which require effective reclamation, it is simply not being carried out. New laws are unlikely to change the situation unless enforcement is in the hands of government officials who love the land more than the coal.

Strip mining as it is now practiced is economically attractive to the coal companies. The increased costs necessary to make underground mining safe cause strip mining to become more attractive than ever. The total cost of strip mining to society, however, is enormous in terms of barren countryside and polluted water supplies. The devastation caused by strip mining has begun to arouse federal lawmakers.[63] But the heart of any law lies in the vigor with which it is enforced. Strip mined coal can be an economical and socially acceptable source of energy for America's future only if reclamation becomes an effective practice where it is possible and stripping is banned where reclamation is not possible.

The alternative to stripping is the underground mining of coal which for many years has been a way of life and death in America. While it may not fit into the familiar listing of environmental concerns, no evaluation of coal as a future source of energy can be complete without facing the health and safety problems of the coal miners.

A first impression must be corrected—underground mining does not necessarily preserve the surface intact. Underground mining of coal seams has resulted in widespread subsidence of the ground above the mine. The subsidence can occur within a few days after mining or as long as fifty years later.

> The land undermined by underground mining alone probably exceeds 7 million acres—with 2 million acres already suffering some subsidence and another two-thirds of a million acres are expected to subside by the year 2000.[64]

The total damage to the surface from underground mining is estimated in the billions of dollars. There seems little that can be done to avoid subsidence if underground mining takes place.[65] Proper land use planning of the surface could, however, minimize future damage.

Underground coal mines are also a major source of water pollution. Cracks in the surface through the mine roof and face permit water to flow through and become acid as it contacts the exposed coal. It then flows or seeps into surrounding water supplies, lakes and streams. The Federal Water Quality Administration has estimated that the adverse effects of mine drainage cost Appalachia $97.5 million a year in recreation income alone.[66] More than 4 million tons of sulphuric acid each year are washed into the waters of Appalachia from coal mines. And this form of pollution continues after a mine is abandoned. The acid mine drainage problem can be somewhat ameliorated by various methods of treatment.[67] Yet, as coal production increases and old mines are abandoned, this significant source of pollution will remain a continuing threat to our objective of cleaning up the nation's waterways.

Underground coal mining also produces "spoil banks" that mar the landscape. The refuse from underground mining and coal preparation is piled up in the mountains of dirt mixed with minerals and low-grade coal.[68] The waste piles grow at a rate of more than 100 million tons a year, not even counting those resulting from strip mines. Some of these piles become slow burning fires that cannot be extinguished and thus are a constant source of air pollution. In Pennsylvania and other Applachian states, there are more than 400 such smoldering refuse banks as well as many underground coal seams burning out of control.[69]

Not only are these waste piles eyesores that burn and pollute the air. Worse yet, they are also sometimes used to form makeshift dams, which back up water into lakes polluted with the wastes from mining operations. These dams generally lack adequate spillways and some are susceptible to breaking in a very heavy rain.

One such break occurred on February 27, 1972, when a dam of the Buffalo Mining Company (owned by the Pittston Company, the largest independent coal mining company in the U.S.) gave way on a mountain just outside Man, West

Virginia. The disaster at Buffalo Creek took 148 lives and left 5,000 people homeless. A wall of water swept through 14 mining communities, destroying most of the houses in its path. The cause of this disaster was simple—human negligence and carelessness. It could have been avoided. The dangers were known both to the U.S. Bureau of Mines and to the Pittston Mining Company. Although one vice president of the company called the disaster "an act of God," local officials in West Virginia were warned in 1967 by the Department of the Interior that the waste pile and 29 others in the state were unstable or could be toppled by high water.[70] Even after this tragedy other makeshift dams have given way.[71] The Buffalo Creek disaster is but another reminder of the coal industry's willingness to risk the ruin of a region and the lives of its people in the process of exploiting coal at the lowest cost.

Coal mining, in short, can do grave damage to the web of life on the face of the earth. But it is the men who go down into the pits who are most directly victimized.

Underground coal mining is the most hazardous occupation in America. A man who has spent his lifetime deep inside coal mines has one chance in twelve of being killed in an accident, the expectation of suffering three or four lost-time injuries, and the great likelihood of incurring "black lung" disease. The accident rate for underground coal miners is the highest of any significant occupation—thirty-five disabling injuries per million man-hours. A coal miner gets badly hurt ten times as often as a steel worker, and three times as often as a construction worker. Over half of the accidents to coal miners are caused by falling rock and coal near the exposed face of the coal seam. Accidents also occur while hauling the coal out of the mine.

The number of miners killed in accidents has been reduced over the years, but mainly because there are fewer miners. Thirty years ago, the industry's rate of fatalities was about one death per million man-hours. A strong coal mine safety law passed in 1969 was largely responsible for the reduction of the fatality rate to 0.66 deaths per million man-hours worked in 1972. Even so, there were still 181 miners killed and while deaths were reduced the accident rate did not significantly improve.[72]

Mining deaths occur almost daily yet they receive public attention only when a dramatic accident kills a large number of miners at once. But the explosions which get the headlines account for only 10 percent of mining fatalities.

The introduction of new machines into underground coal mining has doubled output per man during the past thirty years. By cutting the required manpower in half, the industry has reduced the number of injuries per ton of coal. These new machines—called continuous miners—remove the coal faster, but in the process they increase the level of dust in the confined underground space. As a result lung disease has now reached alarming proportions among miners. The disease caused by the inhalation of coal dust is called pneumoconiosis or more aptly "black lung." It is a continuous irreversible condition that generally debilitates the miner and results in a shortened life span.[73]

The estimates made a few years ago that only 15 percent of coal miners suffered from black lung appear to have materially understated the problem. As a result of a federal program to grant compensation to the victims, medical examinations suggest that as many as 30 percent of the 100,000 active miners have already contracted the disease.[74] The compensation bill alone—which makes no allowance for the pain and suffering of the miner—now promises to be some $500 million a year, more than $1 a ton of coal.[75] As a result, we see that coal has actually been much more expensive to produce than its direct cost to the coal companies. Coal company stockholders along with power consumers enjoying their air-conditioned comfort have received a large and shameful subsidy in the form of the ruined health and early deaths of the miners.

Coal mining will always be more hazardous than selling shoes or playing in a symphony orchestra. Nevertheless, vast improvements can be made. There has been some progress in implementing the new Coal Mine Health and Safety law. The level of coal dust in the mines has been substantially reduced and future generations of miners should escape black lung. A research program is underway to develop safer mining methods, especially near the coal face where most accidents occur. But the enforcement of the new safety standards is still disappointing. Death and injury still haunt the miner. The hazards and destructiveness of coal mining are not inevitable. To reduce them, however, we must face and overcome governmental indifference and corporate insensitivity to human life.

CONVERSION OF THE FOSSIL FUELS TO USEFUL FORMS OF ENERGY

In a sense, our environmental problems only begin with the extraction of energy sources from the earth. Further assaults are made on the physical environment as raw fuel is converted into more usable forms of energy. Of the basic fuels only natural gas and a small share of coal production is used by the ultimate consumer essentially in the form in which it is extracted from the ground. About two-thirds of all fuel produced—crude oil, uranium and most of the coal—are first converted into various different forms of energy.

Energy conversion occurs primarily in "energy factories" —giant oil refineries and electric power plants—which damage the environment in several ways.

Oil refineries convert crude oil, which is more than 40 percent of our total energy supply, into gasoline of various grades for motor vehicles and airplanes, into lubricating oils, into heating oils for homes and commercial establishments and into raw materials for the petrochemical industry. The heaviest residue of the conversion process, the so-called residual oil, is used as fuel for electric power stations as well as other large users of fuel. The crude oil is separated into selected fractions by heating and partial vaporization, converted into smaller molecules, treated to remove sulphur, and finally blended into the desired mix of products.

There are obvious environmental problems involved in the location and operation of an oil refinery. At the moment it is next to impossible to find a location for a new refinery anywhere on the East Coast of the United States. Some of the negative factors are:

- Land-use is a concern because a good site for a refinery with access to water transportation is often a prime recreation area.
- Pollution occurs because large volumes of water are used primarily for cooling, and this water can become contaminated with waste oil before being discharged.
- Solid wastes from coke, organic sludges, and sediments in crude oil must be disposed of on nearby land or in the water.
- Oil spills occur in the area because of heavy tanker traffic.
- Air pollution from the combustion of fuel to supply heat for the refinery can release sulphur compounds and hydrocarbons.
- Odors and noise and other undesirable but expected features of oil refineries; finally, the danger exists of fires and even explosions.

The oil industry has made major investments in recent years to clean up refinery operations. Air pollution control equipment and water treatment and reuse systems are being installed. Nevertheless, an oil refinery remains an environmentally undesirable neighbor in any area where recreation and natural beauty represent a higher use of the land. And it poses a real problem where preservation of the nearby waters and beaches and marine life from oil spills are primary local objectives.

The large electric power station is the other main center where basic fuels are transformed into usable energy. Twenty-five percent of all our fuel is delivered to these stations. In the past twenty years we have doubled and redoubled the size of electric power plants to achieve economics of scale. As the plants grew in size they have been located in rural areas because safety and pollution considerations force their removal from urban sites. But moving to the countryside or even the desert areas of the Southwest has not solved the environmental problems. In any location a huge power plant overwhelms the surrounding environment.

Pollution per kilowatt hour is probably lower in large plants than in small ones. But the size of the large plants and the power lines radiating from them have combined to make power plants a focal point of environmental concerns.[76]

An electric power plant requires a large tract of land—more than 1,000 acres in the case of a fossil fuel plant—because of the need for coal storage and for land to receive the ash and other solid wastes which remain after the coal is burned. The siting of some 500 new plants in the next twenty years—the number required to satisfy current consumption trends—could preempt many of the remaining recreational areas along our bays, lakes and rivers.

The transmission lines that deliver electricity from the power plant to market are also large consumers of land. Existing transmission lines cover nearly 4,000,000 acres of land, an area larger than the state of Connecticut. If current trends persist, during the next twenty years another 3,000,000 acres will be preempted by these more than 100 feet wide rights-of-way with their unsightly towers dominating the landscape.[77]

A major environmental problem arising from electric power plants is the discharge of tremendous quantities of wasted heat into nearby waterways. The average electric power plant converts into electricity only about one-third of the heat energy from its fuel supply. The remaining two-thirds is literally wasted. Existing nuclear plants are less efficient than fossil-fueled plants and emit about 50 percent more heat per KWH of electricity produced.[78] This inefficiency combined with the large size of most new plants results in the concentration of huge quantities of heat at a power plant site. By 1980 electric power plants will require for cooling purposes about one-sixth of the total available fresh water runoff in the entire nation.

What is so harmful about heating the water?[79] As the temperature of water increases, the bio-chemical processes of aquatic life speed up and their need for oxygen increases. However, the heated water can hold less oxygen in solution.

Thus, as the waterway is heated, it is quickly depleted of the oxygen supply needed to support life and to cleanse itself.

A sharp increase in temperature can be quite destructive to fish, although the extent of the fish kill varies with different species. (Power plants using once-through cooling systems also destroy fish, plankton and fingerlings if their screens are not properly designed.) It is also worth remembering that the water being heated is not exactly clean. Many waterways are already near death.[80]

Already overburdened waterways cannot absorb the waste heat from new power plants and maintain water quality standards. The utilities are being forced to build cooling towers or ponds next to the power plants which evaporate most of the heat to the atmosphere.[81] These cooling towers are large—the size of a 30 story building and a block in diameter. They are also expensive, adding some 10 percent to the cost of a plant. And they present environmental problems of their own in addition to their unsightliness. Evaporative cooling means the loss of water, which can be a problem in Western states where water is short. Many cooling towers cause fog and may contaminate an area with salt if the water is at all saline. Thermal pollution can be controlled but the solutions are considerably less than optimum. Dry cooling towers that operate like a radiator and use the air as a cooling medium are possible, but they are much more expensive[82] and are not suitable in all climates.

The problem of thermal pollution should be kept in perspective. Many power plants are spending more money on cooling towers than clean air equipment. Investment in cooling towers should be no larger than what is actually needed to preserve water quality. Further research on the amount of heat waterways can tolerate and on how to dissipate heat more efficiently—or better still, beneficially—could pay big dividends both to the power industry and to the public.

ATOMIC ENERGY

In the debate over atomic power, the public cannot forget the miner and the need to protect him. He initiates the fuel cycle and he is much more exposed to danger from radiation than any member of the general public. Uranium mining is the essential first step in producing atomic energy, and it is a hazardous one. Although uranium can be obtained by strip mining, for the foreseeable future most of it will continue to come from underground mines, where the miners are exposed daily to radiation from the ore.

When exposed to the air, uranium ore gives off radioactive gases that can cause cancer. The critical factor is an individual's exposure over a period of years. The daily dose of radiation in a mine is not immediately dangerous, but each day's exposure has a cumulative effect, which can be deadly. Precisely for that reason, only the future will reveal the full human impact of the rapidly growing occupation of uranium mining which is little more than two decades old. Statistical evidence strongly supports the conclusion that uranium miners have in fact become cancer victims because of their exposure. A study of 3,400 uranium miners revealed seventy cases of lung cancer through 1968, an incidence six times greater than would be expected in such a group. The latency of lung cancer is 10 to 20 years and preliminary information reveals forty additional cases in the study group during the succeeding three years.[83]

It has been suggested that, because most uranium miners also smoke, the statistical studies are therefore inconclusive. Nevertheless, the problem has seemed real enough for the government to move in recent years to impose stricter safety standards in uranium mines. Under these requirements improved ventilation equipment must be installed and other measures taken to remove radiation from the mine. Although these measures are expensive they do not price uranium out of the market. Here again enforcement of the law is the key to its effectiveness. If strict limits on radiation levels are enforced uranium mining can grow without subjecting a new generation of miners to the dangers that the pioneers encountered. But the government's concern for uranium production seems stronger than its zeal for protecting the miner.

Another little known hazard resulting from the mining of uranium is the disposal of some 90 million tons of radioactive sand, the waste products of uranium mining, piled up throughout Colorado, Utah, Wyoming, South Dakota, Arizona, New Mexico, Texas, Washington, and Oregon. Local contractors help themselves to these radioactive "tailings" and use them for such purposes as making concrete blocks for houses. By 1966 the AEC, without publishing any supporting data, concluded that the uranium tailings "present no hazard to the environment, either short term or long term." This conclusion is by no means certain. A study of the birth and death records in the Grand Junction, Colorado, area shows increased genetic problems, higher cancer rates, and lower birth rates than experienced elsewhere in the state.[84] Recently, the State of Colorado and the Atomic Energy Commission signed a contract that will provide for the removal of the uranium waste from beneath homes and buildings.

Atomic power plants have been the center of public debate over using atomic energy. Popular opposition stems largely from fear: Atomic power is still associated in the public mind with the Bomb, a natural consequence of the technology itself and the fact that atomic power plants were developed by the Atomic Energy Commission, the same agency that makes nuclear weapons. Atomic power plants cannot explode, but their operations produce substantial quantities of radioactive contaminants and poisons and thus require an elaborate system of safeguards. Although public fears may be exaggerated, there is no question that nuclear power plants pose serious environmental hazards. For these plants contain the deadliest poisons on earth.

The dangers inherent in the use of atomic energy were, of course, recognized at the beginning of the program.[85] The plants have been designed to prevent radiation from leaking into the atmosphere except in such minute quantities as to be harmless. And the industry's safety record to date in containing radiation has been good. But our experience is relatively limited, and many of the safety devices are untested.

Radiation is a deadly pollutant which if emitted in any sizable quantities could increase the incidence of cancer in the population.[86] The radiation in nuclear reactors is contained by the design of the reactor system which incorporates accident prevention features. These are supplemented by a steel containment vessel which encloses the reactor. Consequently the quantities of radioactivity that leak out of a nuclear plant in routine operation are extremely small.[87] Even the strongest critics of nuclear power concede that the standards now imposed by the AEC are adequate to protect the public from excessive doses of low-level radiation.[88] Actually it is the old X-ray machines in your doctor or dentist's office that still emit excessive amounts of radioactivity. These machines routinely expose individuals to greater radiation dosages than they would encounter living next door to an atomic power plant.[89]

It is fear of an unanticipated accident, rather than routine emissions, that makes nuclear power controversial. A major concern is what would happen if the large and very carefully built piping that delivers cooling water to the reactor were to rupture. If that unlikely event were to occur, the reactor would become so hot that the protective shielding would begin to melt in less than a minute unless emergency cooling water was quickly applied. Atomic power plants are therefore equipped with a carefully engineered emergency system to supply cooling water. Called an emergency core cooling system, it is believed to be satisfactory but has never been put to a test.[90]

"Build now—test later" has been the Atomic Energy Commission's apparent policy for these back-up safety features. Thus far it has been successful. There have been no serious accidents and no deadly emissions of radioactivity from nuclear plants. There is reason to believe that the multiple safety features built into atomic power plants will maintain this record intact, but no one can really be sure. The public can only hope that any defects will be discovered and corrected before an emergency occurs.

Atomic energy is still in its infancy. The oldest commercial atomic power plants are only about fifteen years old and the large 1,000,000 kilowatt units are only now going into service. As of January 1, 1973, there were about 30 commercial atomic generating units in operation with an average life of 5 years. But the nuclear genie has escaped the bottle. An additional 50 large plants are under construction and others are planned. These plants will be expected to operate into the next century. We must consider the hundreds of reactors that will be operating for decades to come. The chance of a major accident, such as a loss of coolant, that could injure hundreds or thousands of people and cause enormous property damage, is probably very small. But we don't know if the risk is small enough to be reasonably confident it will be avoided.

The federal government insures atomic power plants free of charge above a certain minimum. The public has a right to be apprehensive if they are so risky that private companies won't insure them. So long as there is no insurance company willing to bet its money on the safety of atomic power plants, the public has some reason to fear these unknown risks.[91]

Another crucial concern arising from generating electricity from atomic fission is the ultimate disposition of the highly radioactive waste products that accumulate during the normal operation of an atomic plant. These are small in quantity; the waste products from a large power plant would not fill a small room,[92] but they are among the most lethal and long-lived contaminants on earth. If these wastes seeped into any of our life-supporting systems, such as our water, they could be deadly. The most troublesome aspect is that many of these atomic poisons will remain radioactive for thousands of years. At present these highly radioactive wastes are carefully separated when the nuclear fuel is periodically "cleaned" at a reprocessing plant, which is usually located some distance from the power plant. The hot wastes are then stored in tanks buried in the ground at AEC installations at Hanford, Washington, and Savannah River, South Carolina.[93] This is, however, only a temporary expedient and some of the tanks have already begun to leak.[94] The AEC has learned to solidify the wastes[95] and has considered the salt domes in Kansas and other sites for permanent storage. But a satisfactory very long-term solution has yet to be found.

The practice to date in disposing of these wastes has been almost unbelievably irresponsible. Until recently, Great Britain, France and possibly other nations have dumped these highly toxic wastes into the ocean. The ultimate disposition of atomic wastes is perhaps as worrisome as any element of the nuclear puzzle. Man is simply unable to visualize what might happen in the next 25 years, much less in the 25,000-year lifespan of these poisons. Trying to develop a foolproof atomic trash container may be an illusion. Our generation may well be leaving a terrifying legacy of the high energy civilization to endanger the earth for thousands of years to come. "Build now—test later" is not good enough in this case.

The AEC is working on the problem, and it is possible that a safe means of short-term disposal may be found. For now, keeping the wastes in sight, heavily shielded and closely watched, is perhaps the best answer. But a more permanent solution cannot be put off for long. The stark fact is that 146 large nuclear power plants are already in operation, are under construction, or are planned, and all of them will produce these poisons. Only when a permanent disposal system is devised and used will the shadow of catastrophe be lifted from generations whose grandparents are yet unborn.

Still another problem is unique with nuclear power plants—the danger of the spread of atomic weapons. Nuclear power plants produce plutonium as a by-product. Plutonium can be reused in the power plant after it is extracted from the spent fuel at a reprocessing plant. But this same plutonium is also a source material for an atomic bomb. The fact that reactors in atomic power plants routinely manufacture weapons-grade material raises two kinds of fears. One is the fear of proliferation of atomic weapons if nations which are not now nuclear powers divert plutonium from their power plants and make atomic bombs. The other fear is that terrorists or revolutionaries might be able to steal such material and, in a true-life B-movie script, make a home-made bomb with which to blackmail an entire nation. This problem of safeguarding weapons-grade material is deeply rooted in the civilian nuclear technology, and it is so terrifying that it tends to be an "unthinkable" subject, at least for public discussion. With the arrival of the age of the commercial atom, however, the problem must be faced openly.

Proliferation of atomic weapons among the nations of the world has long been a matter of international concern.[96] Despite its assurance of safeguards, the nuclear nonproliferation treaty ratified a few years ago will mean very little unless realistic steps are taken to prevent weapons-grade nuclear material from being diverted. Nations with atomic power plants such as West Germany, Japan, India, Israel, and Sweden are all potential nuclear powers. Actually, any country that develops and operates a nuclear power plant has taken the most important step toward building a bomb. Nuclear weapons technology is now generally known, and the tiny amount of material needed for a single bomb can easily be diverted during the reprocessing of fuel or from plutonium stockpiles. Far greater volumes than are needed for a bomb are already unaccounted for in the United States.[97] And the problem will increase dramatically as more nations build nuclear power plants, especially the nuclear breeder plants of the future, which will greatly enlarge the supply of weapons-grade plutonium produced by the commercial atom.

The United States has by no means neglected this problem. Largely through U.S. leadership the International Atomic Energy Agency has developed a system of safeguards and inspections. Because of the dominant role the U.S. plays in supplying enriched uranium to other nations in the free world, we have been able to insist upon adherence to these safeguards. But assuring the effectiveness of any system of safeguards is a thorny issue that impinges on national sovereignty.

Moreover, the ability of the United States to insist on safeguards because of its near-monopoly of enriched uranium is diminishing. A new method of enriching uranium called the gaseous centrifuge process, is being developed in Western Europe. If it proves successful, this process can be built in a unit small enough for any nation to own and even to operate without being detected. Thus nations will become increasingly self-sufficient in their nuclear programs, and in the absence of effective international control they will be tempted to acquire atomic weapons, in part because of fear of what other nations may be doing.

The other aspect of this problem is somewhat more speculative but even more frightening. It is the prospect noted earlier, that criminals or revolutionaries could steal enough fissionable material to make a home-made atomic bomb.[98] By 1980 there will be literally tons of plutonium in stockpiles, and only a few kilograms is all that's needed for a 20 kiloton bomb. In addition to these stockpiles, the opportunity for theft exists during the shipment of used fuel from power plant to reprocessing plant, which occurs routinely in trucks on the nation's highways.

Hijacking of airplanes has become a frequent occurrence. It is not inconceivable that a revolutionary group or a lunatic might threaten to blow up New York City or Washington, D.C., if his demands were not met. Could such a threat be ignored if there was sufficient reason to believe it could be carried out?

It would be very difficult if not impossible to steal weapons material from an atomic power plant. But the shipments on the highway are more vulnerable. And stored plutonium also could be stolen. The Atomic Energy Commission has by no means ignored the problem but it has kept very quiet about security measures. There is perhaps some merit in the argument that talking about such "unthinkable" problems may exert the power of suggestion, but this can also lead to complacency. Most problems of technology tend to emerge whether we talk about them or not. The public is entitled to know whether we are risking a reign of terror or whether the security system in effect will be adequate in all circumstances.

Atomic power poses a number of risks and troublesome imponderables. Yet it also holds enormous promise for a high-energy civilization that can see the limits of its supplies of fossil fuels. Atomic power does not pollute the air. It is a concentrated form of energy, and it will become even more so if the breeder reactor is perfected. Thus it makes only minimal demands on the earth's surface as compared to strip mining or even deep mining of coal. It offers other potential benefits such as avoidance of the foreign policy and economic problems created by the need to import more and more petroleum.

The fear of atomic power, however, is real and well-founded, and the atom cannot begin to become our society's main energy workhorse until these fears are put to rest. Perhaps only longer experience will allay the public's apprehension. It would be helpful, surely, if atomic power were visibly receiving the kind of candid, tough-minded regulation that answered spoken and "unthinkable" public concerns.

Atomic power was brought into this world under the guidance of a single federal agency—the Atomic Energy Commission. The AEC has been in the curious and often contradictory position of being at once regulator and promoter of civilian atomic energy. Probably more than any other factor, this internal conflict of interest has led to many of the difficulties facing nuclear energy in the U.S. The public record suggests that the AEC has been successful in promoting atomic energy but has failed to convince the public it is a diligent, skeptical regulator. Can the parent really be trusted to judge how well the child is behaving? Separating the promotional and regulatory functions between district agencies is a prerequisite to gaining public confidence and solving the most difficult problems of atomic energy.

AN ENVIRONMENTAL COMPARISON

Perhaps one way to clarify our thinking about energy and the environment is to examine the alternatives now available in the marketplace to supply energy for a use which everyone would agree is essential: adequate heat to keep a house, office, building or factory warm in the winter. A building can be heated with oil, natural gas, or even with coal. It also can be heated with electric power generated by any of these fuels or by atomic energy.

Natural gas is an attractive source of energy: clean, easily transported, and in most areas relatively inexpensive. But there is an acute shortage of natural gas (See Chapter 5). In some areas of the U.S. it is already unavailable for new homes. We must, therefore, look to other sources.

The use of coal would represent a giant step backward, unless technology can be developed to permit burning coal in a large electric power plant without unacceptable air pollution. And there are the basic unresolved questions of how to make strip mining or underground mining of coal environmentally acceptable using present methods.

Oil presents other problems. Burning fuel oil in a household furnace increases urban air pollution even though most of the sulphur is removed at the refinery. Acceptable domestic sources of oil are in short supply. To heat new homes with oil will require expanded drilling offshore where oil spills could ruin prime recreational areas. Increased oil imports, if they are available, will require many more tankers traversing coastal waters. Finally, acceptable sites for refineries to convert crude oil into heating fuel are scarce, especially in areas such as the Northeast where the need is greatest.

Electric heating is, of course, a very clean form of energy use. Pollution of air, water and landscape occurs at the power plant where the electricity is generated and at the mine where the fuel supply is produced. But if electricity is generated in nuclear plants, we face the danger of accidental release of radioactivity and the task of disposing of radioactive waste products in a way fully responsible to the safety of future generations. Electric heat also represents an extremely inefficient use of fuels. In converting the basic fuel to electricity two-thirds of the energy is wasted.

This brief and oversimplified synopsis illustrates our dilemma. No matter which alternative is chosen from among existing energy sources to meet the simple and necessary objective of space heating, society faces the prospect of growing pollution and a diminishing resource base. If one multiplies heating by all of the other necessary and frivolous uses of energy, one begins to get a full appreciation of the energy-environmental dilemma.

Clearly our society needs "something else" as a source of energy—if we can find it. In the meantime, demanding and using less energy may well be our only way to live in a safer, healthier environment.

CHAPTER 5

How Much Energy Is Available?

The motorist who can't buy gasoline is asking some elementary questions: Are we as a nation running out of fuel? How much energy is actually available? Is the shortage a hoax or are we witnessing the beginning of the decline of our high-energy civilization?

We know two things for certain about our energy supply. One is that as far as the consumer is concerned the shortages are real. People cannot buy as much gasoline as they want. Those who want to use natural gas for a new home or factory are unable to buy it. Many industries have seen production curtailed because they couldn't obtain sufficient fuel in any form. But we also know for certain that billions of tons of coal and billions of barrels of oil and trillions of cubic feet of natural gas are buried somewhere in the ground. The inescapable conclusion is that the natural storehouse of energy is far from being exhausted even though we cannot completely dismiss the neo-Malthusian specter of one day actually running out of energy.

To be sure, we face a very serious problem—the need to supply an ever increasing growth in demand for energy. The

U.S. consumption of energy is immense, the equivalent of 13 billion barrels of oil in 1973. Annual increments of growth approach the equivalent of 500 million barrels of oil. The supply problem is to surmount the man-made and natural obstacles that stand in the way of increasing production by such vast quantities each year.

We are not about to run out of energy but we have been unable to find it, bring it out of the ground, and convert it into gasoline or electricity as fast as people buy more cars and air conditioners. The energy industry's sales department has met with far greater success than the production department. The reasons for the demand-supply imbalance are at the heart of the conflicts over U.S. energy policy.

Any assessment of U.S. energy supply must include the rest of the world. For the U.S. already imports some 30 percent of its oil supplies and much of its increased supply of natural gas. At the same time, the U.S. is the major source of enriched uranium for Western Europe and Japan. The U.S. is an integral part of a worldwide energy network. In fact, most of the large U.S. petroleum companies are engaged in discoveries and production of oil and natural gas throughout the world. Thus in appraising U.S. fuel supplies we must also examine world sources of supply as well as demand for these supplies by other advanced nations.

Energy supply is usually discussed in terms of the quantities of discovered fuel remaining in the ground, called "reserves," or the ultimate size of the resource base referred to as "resources." We are, of course, interested in such estimates because they tell us something about the life-span of fossil fuels and when our resource base may peak out. But the fact that the oil is in the ground doesn't mean motorists can buy as much gas as they need. Far more pertinent as a cause of the energy shortages in the next decade and beyond are the limits on our ability to increase the volumes that can be produced each year. It is the productive capacity—the amount that can be delivered to each home or car each day —that is the key figure rather than the reserves or resources in the ground which might just as well be mud if the capacity and incentive to produce and sell it do not exist.

The key question is how quickly fuels can be brought to market in the forms consumers need. How quickly can we expand the pumping capacity, the mining capacity, the refining capacity, the electric power plant, the transmission lines, and the rest of the energy infrastructure? Can all of these necessary elements of supply be expanded as rapidly as an energy-hungry America indulges its apparently insatiable appetite for fuel?

During the 1960's we were lulled into neglecting the serious difficulties involved in expanding the productive capacity of U.S. energy supply. Demand expanded but the nation had vast surpluses in productive capacity. At the beginning of the decade, oil- and gas-well production as well as coal mining capacity were greatly in excess of our annual consumption of fuels. Each year consumption increased. In fact, during the 12-year period 1960 through 1972, energy consumption increased 60 percent. But we did not increase our capacity to extract fuels from the earth, or to convert them into gasoline, fuel oil and electricity.

In effect, we "ate off the shelf" in 1960's and now, midway through the 1970's, we find the shelves are bare. Consumption grows each year but our ability to expand fuel production in the U.S. has virtually disappeared. Oil and natural gas form 75 percent of the total energy supply, and we are pumping all of the available oil and gas reservoirs at mmaximum efficient capacity. But the sad fact is that production in 1973 was no greater than in 1970.

The question immediately arises: If there is plenty of oil, natural gas and coal in the ground or near the surface, why have we failed to do whatever is necessary to tap it and convert it into usable forms for the consumer? In the face of actual shortages, why is the nation unable to achieve the increases in production necessary to fill the gap? The answer is that there are limits and constraints on the ability of energy companies to expand production, particularly by the massive amounts required.

The U.S. hasn't run out of energy but its productive capacity shows every sign of peaking out. There are real limits to the rate of growth in energy production. Some of these constraints are physical and economic in nature; others express basic values that conflict with energy production and reflect our democratic system of government which tends to slow down the decision-making process to afford due process to accommodate all conflicting interests. These limits and constraints on increased production are not likely to disappear or diminish in the foreseeable future. The constraints turn on such words as "price," "environment," "public domain," "lead time," "resource base" and "politics."

Within the U.S., the pace of exploration and development for oil and gas and the construction of new oil refineries, coal mines and other links in the energy supply chain are all influenced by the price the consumer pays. In recent years, prices of fuels have been more or less controlled by government to combat inflation. Consumers and government alike might well suppose that profits are sufficiently high to encourage an all-out search for more supply by the energy companies. Certainly recent price increases should spur the search for more fuel. But for many years industry has taken the position that the price for natural gas, crude oil and even coal has not been high enough to warrant major expansion of the search for new sources or the construction of new mines.

One might describe the situation as a sit-down strike by the producers, at least within the U.S. But a basic fact needs to be emphasized: The U.S. companies in the business of producing energy fuels are not utilities, and they are under no legal obligation to drill or dig for fuel. If they do not believe the price of new supplies is right, they can leave their capital in the bank and the fuel stays in the ground.

Price does, in fact, determine how much energy in the ground is potentially available as well as the incentive actually to produce it. As the price rises, it obviously becomes more attractive to produce fuels that are more costly to find and extract. There is a big difference between the geologists' version of our national resource base and the base that is economical to produce. Much of the remaining fossil fuel in the U.S. may be as costly as the gold in the ocean which is so well diluted that even at prices above $100 an ounce it doesn't pay to recover it.

A related consideration is that brighter petroleum drilling prospects abroad compete for investment with expanded drilling in the U.S. The large petroleum companies that explore throughout the world are likely to put exploration capital where the company's geologists say they can find the most oil for their money. Nowadays that is often outside the U.S.

The recent revolutionary increases in the price of oil in the world market may change the picture dramatically. Prices have risen so high that it is difficult to believe that they could be a constraint in finding new sources of fuel with the exception of natural gas, which is still regulated. Yet price controls are an important part of federal policy to control inflation and prevent windfall profits to energy producers.

In the preceding chapter we reviewed the environmental constraints on the production of energy. These place stringent limits on the rate at which energy fuels can be extracted from the earth no matter how high the price may go. And for good reason. An example is the risk of oil spills which increase dramatically if one tries greatly to accelerate the pace at which off-shore drilling is expanded. The government knows little about the off-shore areas and employs relatively few experts who can discover new information. The danger in a rapid and uninformed acceleration of off-shore drilling is compounded by the fact that the industry would be stretching the limits of its trained manpower. And the government would be stretching its ability to inspect the operations and enforce environmental regulations. For these reasons the off-shore drilling program for oil and natural gas was suspended for almost two years after the Santa Barbara oil spill and the subsequent expansion of the program has been at a slow pace, despite admonitions for greater speed in two presidential energy messages.

What is true of off-shore drilling for petroleum is even more true of coal production where environmental constraints on growth have brought expansion to a virtual halt. Strip mining of coal must now meet reclamation requirements and deep mining must comply with new safety requirements that inhibit construction of new mines. Even more pressing are air quality laws. The technology to burn most coal in an environmentally acceptable manner has been slow to emerge and thus the bulk of coal supply has been outlawed and expansion of production severely limited.

Environmental concern affects not only the pace of fuel production but also the rate at which oil refineries, power plants, oil pipelines, electric transmission lines and every phase of the energy supply can be expanded. It is a new fact of corporate life that expansion of production in an environmentally acceptable manner involves more time for planning and for public participation in the decision-making process. Production need not stop, but the need for environmental review slows the pace at which energy companies can expand production.

The limits on expansion of energy supply are actually more fundamental than public concern about air or water pollution. The slower growth movement is perhaps the fastest growing popular movement in America. Proposals to industrialize undeveloped areas are meeting with intense opposition. Thus almost any project to mine fuel, build power plants, refineries or any energy producing facilities meets intense opposition from citizens in the region who don't want their lives changed. This mounting opposition to development is a growing constraint on the pace of expanding energy production.

Still another constraint, and a severe one, is that most of the remaining fuels are located in the public domain and are thus owned by the government. Private industry must obtain government permission before it can exploit fuels on public lands. That permission should not be given lightly. In the off-shore areas government has behaved much like a sensible monopolist. It has sold the fuel it owns at a pace slow enough to attract large bids from industry for what is offered. The pace pleases the environmentalists and the government budget-makers, who are a powerful combination.

The stakes are large. Mineral lease sales, almost entirely gas and oil lands, brought some $4.2 billion into the Federal Treasury from 1966 through 1971.

But pressure is now intense for throwing open the federal domain. In his energy message of January 23, 1974, President Nixon ordered a ten-fold increase in the off-shore leasing effort by 1975. Past experience suggests that such an expansion cannot be achieved without a gigantic giveaway of public property and sacrifice of environmental values. It will require a great expansion of the government's capability to evaluate the resources it is leasing, to assess potential damage to the environment, and actually to conduct the sale of leases. The National Environmental Policy Act requires the federal government to evaluate the impact on the environment before it makes such sales, and environmentalists have obtained court orders to stop prior sales that failed to comply.

Expanding the government's capability to lease what it owns points up a major problem today in evaluating how much energy might be made available to meet future needs. The government actually knows very little about its vast storehouse of oil and natural gas in the outer continental shelf or its coal lands in the West. The data available to the government on the cost of production from existing leases is kept under wraps and no effort is made to compile and publish it in a meaningful way. The energy companies provide little information on production costs. There is a serious danger that a speed-up in leasing would cause the government to sell without knowing the value of rights or whether the public is receiving a fair return. A balanced leasing program thus automatically limits the rate of growth in the production of energy. And the industry's ability to expand production is constrained by the necessarily slow pace of government leasing.

Even if we created an energy company's dream-world—high prices, no environmental-protection laws, and a government leasing policy of "stake a claim and it's yours"—there still would be a constraint on the rate of growth in energy production, especially petroleum. The remaining oil and gas reservoirs in the U.S. are not exactly "bubbling out of the sand" like the oil in Saudi Arabia, and in any event have never rivalled the Middle East's productivity. These U.S. reservoirs must be discovered and developed, which requires a lead time of perhaps four years for each reservoir. The lead time lengthens because the easy "finds" have already been discovered. Remaining petroleum reservoirs are smaller in size, deeper in the ground, or farther out to sea, all of which require a longer lead time before production begins. Except for Alaska's North Slope, the last really large field found in the United States was the Louisiana West Delta discovery in 1962.

To be sure, if the industry tripled or quadrupled its drilling effort more of the remaining reservoirs could be found sooner, but there are inherent limits on how quickly the drilling effort can be expanded. Even with every incentive, it still takes time to build and man new drilling rigs. There are shortages of tubular steel for drilling rigs. Men practiced in the dangerous art of off-shore drilling, whose skills cannot be acquired overnight, are few and in great demand.

As for coal, we know where it is located in abundance but there is still a four or five year lead-time in designing and constructing a new coal mine. And the same limitations of manpower and materials determine how many new mines might be simultaneously opened and put into production.

The shortage of skilled manpower limits the rate of growth throughout the entire energy supply system. To expand deep mining of coal dramatically would require attracting young people to an occupation that at present has the worst safety record in the nation. The building of power plants and refineries is already plagued by shortage of skilled construction workers. These shortages account for many of the delays and breakdowns in the power plants already built. Atomic power especially cannot be safely expanded without highly-trained personnel. The rate of growth in production is thus directly constrained by the shortage of skilled people and the slow growth in their numbers.

The size of the resource base is also a constraint on the rate at which domestic natural gas and oil fuel production can be expanded. The U.S. is a rather well-developed petroleum province. There may be more petroleum left in the ground than has been produced, but it is more difficult to find. Many discoveries are needed merely to replace reservoirs already discovered that will soon be depleted. Each new discovery reduces "what's left" and thus limits the rate of expansion. The resource base is thus limiting the rate of growth in oil and natural gas production in the U.S. We are "peaking out" even though we are far from running out.

The limits we have identified on the production of fuels apply to all the steps in energy manufacturing. The much-delayed Alaska pipeline provides a vivid example. Oil refineries, uranium enrichment plants and power plants to convert the raw fuel into usable form—these too face the same constraints stemming from a ceiling on profits, environmental opposition, shortages of skilled manpower and the uncertain timing of government decisions. There is also a limit as to how fast tankers and pipelines can be built to transport oil. Any step in the chain of supply can become a bottleneck that will cause shortages of energy in the forms needed by consumers.

All of the constraints find their expression in the world of "politics." The rate of growth in energy production is in conflict with values strongly held by major segments of our society. Consumers tend to believe that the energy companies are making enough profit and don't need large price increases to spur more rapid exploration. The environmentalists tend to feel that limits on production are essential to prevent destruction of the ecology. In a democratic political system, these values are bound to make themselves felt. This has been especially true in recent years as the consumer movement gained strength and the general trend toward participatory democracy gained wide acceptance. The political process always accommodates countervailing forces and then strikes a compromise. Thus, when President Nixon in his 1973 Energy Message sought to suggest measures to "increase domestic production of all forms of energy," he found it necessary at the same time to acknowledge the goal of "the lowest cost consistent with the protection of both the national security and the natural environment."

The energy producers have run into conflicts over social goals so fundamental that they represent real obstacles to growth in production. No matter how intensely the nation wants to increase energy production, it is not prepared to surrender its anti-inflation and environmental protection goals. In addition, there are the lead-time problems as well as the limits of the resource base itself. All things considered, energy production in the U.S. from existing sources will do well to show even small rates of growth in the years ahead.

The root of the energy problem is that in recent years there has been no limit on growth in demand, but production from existing sources has come under mounting pressure. Each year we fall farther and farther behind in meeting demand from domestic production, thus creating shortages and a wider gap to be filled with imports.

In the winter of 1973-74, the world experienced an object lesson in the political constraints that exist on oil supply. The Middle East still contains an abundance of petroleum in relation to demand, but the political objectives and economic incentives of the Arab producing nations have led to stringent limits on how much is produced. Here again oil in the ground is not the problem; it is that many producing nations with much oil and few people are simply not interested in producing in accordance with the standard western economic textbook model or urgent western demands.

It is not simply the Arab nations that are determined to be out-of-step with the demands of energy consumers. Such nations as Canada and Norway have advanced good reasons for exploiting their energy resources at slower rates of growth than is physically and even environmentally feasible. Canada is concerned about distorting its economy through excessive raw materials exports and causing inflation from the flow of necessary investments. Norway faces a political decision on how to spend or invest surplus money if it accelerates its North Sea development, a problem its leaders apparently want to avoid.

Energy supply is indeed a complicated, inter-disciplinary problem, yet most discussions treat it in an amazingly simplistic manner. Intelligent economists say flatly that the shortage of natural gas will be cured if price controls are removed and that higher prices will bring a surplus of oil to market in a few short years. They are blind to the environmental limits imposed on drilling off shore as well as the environmental and political opposition to development in the U.S. and around the world. Geologists describe the gigantic "supply" of energy fuels in the ground without considering how much can be produced for the equivalent of $4, $10 or $100 a barrel. Consumers suspect a conspiracy because energy is in short supply while they read geologists' reports stating that we haven't run out of fuel in the ground. And environmentalists suggest using cleaner sources of energy that exist only on paper without any appreciation of the capital investment and lead times required to "harness the sun."

The familiar analysis of our fuel supply stressing "barrels of oil" in the ground encourages a false sense of security and ignores the crux of the supply problem. This planet is not running out of fuel. The difficulty is that with existing sources there is a limit on how quickly we can extract what's in the ground and deliver it to the consumer in usable form. That limit is already being felt and will determine how much energy is available in the decades ahead.

Recognizing the limits on expansion of supply, it will nevertheless be useful to examine statistics on conventional fuel remaining in the earth's surface. In making our appraisal we should be aware that until recently all U.S. experts on energy assured us that we had ample resources of coal, oil and other fuels at or near current costs to last for the rest of this century or longer. Typical of such pronouncements was a massive study by the many interested departments of the federal government published in September of 1966, which found that:

> In light of present day technology, the nation's total energy resources seem adequate to satisfy expected energy requirements through the remainder of this century at costs near present level, but technological advances will be required to reduce costs and extend the supply base into the more distant future.[99]

These pronouncements of abundance failed to consider how difficult it is to expand production to keep pace with exponential growth in demand in the face of multiplying constraints. Such empty reassurances also contributed greatly to the shock that Americans experienced as shortages later developed.

We shall examine each of the significant sources of energy supply currently in use.[100] It should be kept in mind that to a considerable extent these sources are interchangeable over time. For example, we can heat a new home with oil, natural gas, electricity generated by coal or atomic energy. But once a home is built with a natural gas furnace, it is very expensive to switch. We should also recognize that many end-uses, such as the automobile, are now dependent on a single form of energy. Switching fuel would mean building an entirely new electrified transportation system.

Thus while a shortage in one fuel can potentially be filled by another, in the next decade and beyond, a shortage in one area is more likely to have a domino effect. For example, a gas shortage might cause people to switch to oil and thereby create a general fuel shortage, such as we experienced in the winter of 1972-73.

OIL—AN UNCERTAIN FUTURE SUPPLY

Americans consume more oil than any other source of energy, and our demand has been increasing each year. But much uncertainty hangs over the future of oil supplies in the U.S.

Unlike some of the other highly industrialized nations—Japan and most of Western Europe, for example—the U.S. is fortunate that major oil reserves lie beneath our soil and adjacent offshore waters. We have been the world's leading producer of oil for many decades.[101] From the late nineteenth century to the late 1960's, our booming domestic oil industry drilled more than two million holes into U.S. soil. As a result we have discovered and produced a large share of the total oil resources of the U.S.

From the first commercial well in 1859 through 1972,

production of crude oil in the U.S. has totalled about 103 billion barrels. More than one-third of this total was produced within the past dozen years alone. Production is thus increasing at a breath-taking pace against a resource that is fixed in ultimate quantity.

One wonders then how much oil is left and at what rate it can be discovered and produced. In early 1973 there were about 45 billion barrels of proven oil reserves including the untapped finds on the North Slope of Alaska.[102]

These reserves, except for Alaskan oil, are being pumped at their maximum efficient rate and therefore cannot provide the source for expanding production. In fact, without new discoveries, current production would taper off to next to nothing in a decade. But proven reserves include only oil that is already found and known to be economical at current prices. These reserves are in reality a working inventory and are by no means the measure of how much oil remains in the ground.

The proven reserve figure can be adjusted up or down by changes in technology and the price of oil as well as by improved knowledge of the size and location of the reservoir. For example, the 45 billion barrels of proven reserves represent only about 30 percent of the oil in the known reservoirs because current technology and prices permit recovery of only about 30 percent of the oil in a given field.[103]

Oil in a reservoir does not occur in a free-flowing pool. It is found in porous rocks or trapped under high pressure in dense rock. An increase in the percentage recovered, therefore, cannot automatically occur. It requires additional expenditures for secondary and tertiary recovery techniques that entail a higher production cost per barrel, and subsequently would require a higher price or subsidy for the extra oil discovered.

Of more fundamental interest are the statistics encompassing the amount of oil in reservoirs that still remains to be discovered in the U.S. Here we become engaged in a numbers game of major proportions—a game which we play almost blind-folded for lack of knowledge as to costs and environmental acceptability of the resources. The range of error in these estimates is very wide. The upper limits are suggested by the U.S. Geological Survey. It estimates that the amount of oil in place in the United States and its continental shelves is probably close to 2,000 billion barrels. This figure is, however, completely unrelated to any concept of costs and could be most misleading.

In sharp contrast, conservative authorities such as M. K. Hubbert believe that the total petroleum resource ultimately recoverable in the U.S. is no more than 200 billion barrels, which includes the 103 billion barrels already produced.[104] Other estimates range from 300 to 400 billion barrels recoverable with present technology and within the range of today's prices,[105] to an ultimate find of 600 billion barrels if higher prices and environmental concerns permit the recovery of more marginal resources.[106]

If oil consumption continues its recent rapid rise, no fewer than 250 billion barrels will be needed in this country in the next 30 years. More astoundingly, if growth in demand continues unchecked, we may require 600 billion barrels more in the 30 years that follow. Such growth in oil consumption could completely drain the conventional oil supply in the U.S. within the next 50 years. Even at reduced rates of growth the U.S. could run out of oil in 100 years. These numbers, however, provide a false sense of comfort; as the motorist knows, the problem of an empty gas tank is not 50 years away.

The more immediate concern is the limits on the rate at which this oil can be found and produced. Production will peak out long before the resource base is exhausted. Oil production in the U.S. may be reaching its peak and the end of its growth. All existing reservoirs tend to produce less oil each year. Thus more and more new discoveries are needed simply to maintain current production rates. Even assuming more optimistic resource estimates, conventional oil production in the U.S. will level off in a decade or two and certainly before the end of this century. The more pessimistic view is that the nation is already at or very near its peak in oil production. To boost oil production, the industry would need a drastically stepped-up program of exploration and development, which would require a dramatic reversal of current trends. In view of the constraints of price, environment, government leasing, and politics discussed earlier, this seems highly unlikely.

A look at current drilling trends for oil and natural gas gives a fairly accurate indication of the scale of expansion required. In 1972, total exploratory drilling in the U.S. was 45 million feet, a sharp fall from the record level of 74 million feet in 1956 and the level of drilling throughout the 1960's.[107]

Actually the number of new exploratory wells fell even more sharply because producers were forced to drill deeper for new discoveries. The volume of oil resulting from each foot of hole drilled has dropped. Thus, while demand is up, drilling has gone down and discoveries of oil are down even more. For almost two decades the industry has been tapering off the search for new oil fields in the U.S., and the reason appears to be that the future oil prospects offer relatively slim pickings.

The pace of drilling is of course now increasing in response to price increases. Even so it seems unlikely that the industry will launch the kind of dramatic increase in drilling that would appreciably expand domestic oil production. Any increases at all are three to five years away because of the lead-time required to bring new oil fields into commercial production.

It is estimated (See Table 1) that more than 50 percent of remaining discoverable oil is located off-shore and in Alaska. The offshore areas that appear most attractive to the industry for oil drilling lie under the waters off the coasts of California, Texas, Louisiana, Florida, New England, and the Middle Atlantic states, and in the Gulf of Alaska. The Nixon Administration is committed to moving ahead with offshore drilling but the citizens of Florida, California, New England and Long Island seem equally determined to keep refineries, oil rigs, and the threats of spills away from their shores.

Table 1

OIL IN-PLACE[108] RESOURCES OF THE UNITED STATES

(billions of barrels)

Region	Ultimate discoverable oil	Reserves discovered to 1/1/71	Remaining discoverable oil	% of ultimate
Lower 48 states				
Onshore	561.8	384.9	176.9	31.5
Offshore and South Alaska	128.6	16.3	112.3	87.3
Alaskan North Slope				
Onshore	72.1	24.0	48.1	66.7
Offshore	47.9	———	47.9	100.0
TOTAL UNITED STATES	810.4	425.2	385.2	47.5

Thus a cloud of uncertainty hangs over the future of domestic oil production. For the past three years the industry has not even been able to bring new oil into production as fast as existing fields have been depleted. All of the limits on the rate of expansion discussed above are concentrated on oil. There is a serious question whether the U.S. resources can be discovered and produced quickly enough to support a significant increase in production. To the extent production is increased the limited U.S. supply of conventional oil will be exhausted that much sooner. The most realistic expectation is that domestic oil production has reached its peak.

NATURAL GAS—CLEAN AND DESIRABLE, SHORT ON SUPPLY

Next to oil, natural gas is the nation's largest primary source of energy, accounting for more than 30 percent of current supply. Natural gas has satisfied the bulk of energy growth in recent years, supplying more than half the rising demand for fuels in the 1950's and 1960's.[109]

Recently the gas supply picture has changed. It came as quite a shock when local gas utilities hung up "Not for Sale" signs to potential new customers and frequently cut off the gas to some old customers. The country has entered a period of natural gas shortage which may never be reversed. Natural gas is running short at a time when it is more in demand than ever. The advantage of gas as a clean, environmentally acceptable fuel has made it our most desirable source of energy, and it has also made it scarce.

The shortage of natural gas and oil developed so rapidly that many people question whether it is real. Some consumer groups are convinced that the gas shortage is a fake, especially because prices are still controlled by the Federal Power Commission and the industry is using the shortage to press for decontrol. Consumers suspect a conspiracy by gas producers who are sitting on their wells hoping to force the public into paying higher prices.[110]

Actually no one can be certain that some gas fields are not shut-in or that discovered gas reservoirs are not being developed now in hopes of higher profits later. The information available comes primarily from the industry. However, even if one assumes that some companies are playing a waiting game, there is still a tremendous gap between the demand for natural gas and the potential supply.

According to the industry—and just as with oil there is no other source of data on reserves currently available—present discovered reserves of natural gas in the U.S., including Alaska, total about 268 trillion cubic feet. Thus mathematically, present gas reserves would only meet current annual demands of some 23 trillion cubic feet for about 11 years with no expansion to meet growth. Actually the discovered reserves would not sustain even the current level of production beyond a decade. Gas wells, not unlike people, tend to produce less as they grow older. Major new discoveries are needed to maintain natural gas production at current levels, much less meet the huge expansion in projected demand. The conventional estimate calls for more than 700 trillion cubic feet of gas between now and 1990. If this trend persists, an additional 1,400 trillion cubic feet would be needed in the subsequent 20 years.

Estimates—again only those made by the industry are available—vary as to the remaining gas resources, the gas not yet discovered. This is estimated at from about 1,200 trillion cubic feet[111] to an ultimate potential resource of 2,650 trillion cubic feet. As with oil, gas supplies available to the consumer will depend on the amount of gas actually recoverable and the speed with which supplies can be found and produced.

In contrast to oil, natural gas remaining is believed to be twice as much onshore as offshore. The natural gas onshore is mainly in deep formations (15,000 feet or more below the surface) and in tightly trapped reservoirs in the Rocky Mountains of Colorado. Here is one breakdown of remaining supplies:

Table 2

RECOVERABLE GAS SUPPLY[114]

(trillion cu. ft.)

Region	Ultimate gas discoverable	Gas discovered to 1/1/71	Gas remaining	% of ultimate
Lower 48 states				
Onshore	963.1	413.5	550.0	59
Offshore	260.1	45.9	214.2	82
Alaska	227.4	26.0	201.4	88
TOTAL	1,450.6	485.0	965.6	65

These natural gas resources will be more expensive than present supplies because they come from farther below the earth's surface and from the northern frontier areas of the continent. Finding these supplies and transporting them to market will take longer and will require more sophisticated equipment than ever before. In addition, environmental concern over the drilling and transportation of the associated oil may well delay the development of off-shore gas fields.

In order to satisfy the potential for growth we would need to discover some 30 trillion cubic feet of natural gas each year for the next 10 years. The average annual discovery rate for the past decade has been 17 trillion cubic feet, and even that level has dropped to 11 trillion since 1968, excluding Alaska.[113] The U.S. has been drawing down its gas reserves every year since 1967. To meet projected needs in the next decade, the current pace of drilling for natural gas would have to be tripled and the rate of discovery markedly improved.

There is a serious question whether the necessary tripling or quadrupling of the pace of drilling activity in the U.S. is a realistic possibility. Most of the natural gas offshore in associated with oil and so is the drilling. Environmental concerns impose serious constraints along with the numerous other limits on the rate of expansion previously discussed. Even if the industry launched an all-out drilling program in face of these disincentives, the enlarged discovery levels probably could only be maintained for a few years.

Natural gas is our most limited domestic resource. Even if the most optimistic forecasts of a supply-demand response to a free market were fulfilled, enlarged production levels would surely peak out before 1990. Thereafter, levels of production could not be increased and would taper off no matter how high the price of gas or how many holes were drilled because the resource base would limit supply. The more pessimistic view is that we have already reached that limit.

Though it may be difficult to believe, natural gas production in the U.S. may be at, or very near the peak rate which the resource base can supply. In fact we are already forced to import natural gas at much higher prices than the domestic price to meet consumer requirements. Imports from Canada and from overseas—special tankers transport gas in liquified form—are filling part of the gap.[114] We are also converting oil into gas and trying to learn how to gasify coal economically.

COAL—SCARCITY AMIDST ABUNDANCE

More than any other fuel, coal illustrates the frustrating dilemma we face with our current energy supplies. At present coal is our most abundant source of energy, but we cannot mine it or burn it in a socially acceptable way.

The size of the U.S. coal resource is roughly thirty times the energy value of our oil and gas inventories combined. Our knowledge of the U.S. coal resource is inadequate, but we know it is so large that we should not be in danger of running out of coal for at least another century. According to the estimate of the U.S. Geological Survey, based on mapping of known fields and exploratory data, remaining coal deposits in the U.S. total approximately 1.56 trillion short tons.[115] Some 200 billion tons of recoverable coal lie within 1,000 feet of the surface in seams at least 3-1/2 feet thick.[116] Even this conservative figure of 200 billion tons is huge when compared to a domestic consumption of about 600 million tons last year. It is evident that, in physical terms, a doubling or even tripling of coal consumption could be sustained far into the next century. It must be emphasized, however, that these are quantities of coal in the ground and are not proven reserves. The coal industry, unlike the oil and gas industry, does not maintain a nationwide inventory of discovered reserves, and there are no figures for coal comparable to the discovered reserves for oil and gas.

The truth of the matter is that no one knows who owns America's coal resources or how much is really available at any given price. Just as in oil and gas there are no supply curves. The figures on total coal resources, however, are fairly reliable because estimates are based on outcrops and other tangible evidence.

There are many obstacles in the way of the rapid development of this abundant resource. As noted in the previous chapter, the underground mining of coal is the most dangerous occupation in America today. The Coal Mine Health and Safety Act of 1970 attempted to remedy these abuses, but in doing so it has added considerably to the costs of underground mining and, therefore, to the state of uncertainty already existing as to future supply.

For these reasons strip mining has become much more economically attractive and has been steadily increasing its share of the coal market, producing some 258 million tons in 1973, up from 165 million in 1965. Strip mining now accounts for about 50 percent of current production. The strippable coal represents only some 10 percent of the total coal resource but it may well represent a much higher share of the U.S.'s economically recoverable coal.[117]

The pace at which we make use of our coal resources will be determined in large measure by our ability to develop technology and skills to mine, process and consume coal cleanly. But the shortage of skilled labor and long lead times for opening new mines are not to be ignored. The location of the remaining coal resources is another factor that will influence future use of the fuel. The overwhelming bulk of the coal until now has come from Pennsylvania, West Virginia, Illinois, and, to a lesser degree, Kentucky. In the last twenty years these states have accounted for 70 percent of U.S. production.[118]

The remaining reserves of these eastern states—primarily deepmine coal—are still quite substantial, yet it is important to know that there is more coal west of the Mississippi River than to the east and that the importance of western coal will grow in the years to come. The large western coal fields are low in sulphur content and are currently attractive because their coal complies with air pollution standards. But they are lower in heating value than eastern coals and this disadvantage is compounded by the fact that they are located far away from the major consuming regions.

Coal is expensive to transport. In order to comply with air quality standards, some utilities are bringing coal 500 to 1,000 miles from the Rockies to the industrialized east, thus paying double the delivered cost of coal from nearby eastern mines.

The importance of western coal will nevertheless grow in coming years. Having lost many markets to gas and oil, coal will soon outlive these more scarce resources and can regain

Table 3

COAL RESERVES[121]
(billion tons)

State	Bituminous	Sub-bituminous	Lignite	TOTAL
Colorado	62.4	18.3	——	80.7
Montana	2.3	131.9	87.5	221.7
New Mexico	10.8	50.7	——	61.5
North Dakota	——	——	350.7	350.7
Utah	32.01	0.2	——	32.3
Wyoming	12.7	108.0	——	120.7

these markets through technology that converts coal to synthetic oil and gas. Coal-gas or coal-oil plants located near tthe mines could then pipe synthetic gas or oil to consumers in other parts of the country much more economically than transporting the coal by rail. Perhaps of even greater long-range importance, the coal reserves in the east and the west are a unique combination of hydrocarbons which, not unlike petroleum, can serve as raw material for the rapidly growing petrochemical industry. But these uses must await new technology.

An important additional factor contributing to the uncertainty surrounding the availability of coal supplies is the ownership of our coal resources. Much of the remaining coal resource in the east is owned privately, and many of the most promising tracts are in the hands of the steel industry and other private interests holding them for future development. Recently oil companies have acquired billions of tons of coal reserves in the east and the west as well.

Coal is America's resource giant in a straitjacket. The shortage of natural gas and the slumping domestic oil industry have created an energy vacuum which coal could easily fill. But coal, the filthy villain of our industrial era, is still part of the problem instead of the solution.

ATOMIC ENERGY—A MIGHTY QUESTION MARK

Shocking as it may seem, we face a shortage of atomic energy even before this new entrant on the energy scene makes much of a contribution. Today's atomic power plants exploit only about two percent of the energy potential of the uranium used. The size of the U.S.'s uranium resource is not small but at a 98 percent rate of wasteful consumption even a large resource will not last long. Unless more efficient atomic power plants are soon developed, atomic energy may be only a passing phenomenon rather than the wave of the future.

Because of the AEC's long involvement in uranium purchases and its concern over supply, we have rather good data on the size of this energy resource. The discovered reserves of uranium in the U.S. today amount to about 250,000 tons at the current price of $8 a pound, or 390,000 tons at $10 a pound.[120] These reserves have been growing in recent years and are adequate to meet estimated requirements for 10-15 years. The current market is soft and a government stockpile of 50,000 tons hangs over it.

Thus the uranium supply picture in the United States is not one of present shortage, as in natural gas, or even of tight supply, as for oil. But the current situation is going to change rapidly.

Consumption is sharply rising. In 1973 deliveries of uranium to the electric power industry totalled some 15,000 tons, equal to the present annual production of U.S. uranium mines. By 1980 demand for uranium will almost certainly double and redouble again by 1985. Thus in little more than a decade, if atomic power is to play its expected role, America will probably be consuming uranium ore at the rate of 60,000 tons each year. Demand is soaring as atomic energy begins to supply most of the growth in electric power production. And even this amazing rate of growth assumes that nuclear energy will supply only about 10 percent of U.S. energy needs by 1985.

Official AEC estimates are that the total uranium resources in the U.S. at $10 a pound are only about one million tons and at $30 a pound some two million tons.[121] The real crunch on available uranium supply will come in the 1980's. We will need a fourfold increase in capacity to produce the necessary uranium in a little more than 10 years. This raises a question as to the industry's ability to expand fast enough to meet the demands of the market.

The amount of uranium in the ground is not likely to be the limiting factor, certainly in the next two decades and perhaps for a longer period. But there are serious constraints that place a cloud over the rapid expansion of nuclear power in the 1980's. Exploration for additional uranium is inhibited because the government has a stockpile of 50,000 tons that it acquired to keep the industry alive during prior decades. Uncertainty about when the ban on imports might be lifted also discourages investments in new mines. Uncertainty over price combined with the lead time in expanding uranium mines raise questions as to the industry's ability to expand at the soaring pace that may be required. The problems of using leaner uranium ores are not adequately reflected by mere estimates of the higher cost of mining.

In its most concentrated form uranium is only one-fifth of one percent of the mineral deposit. One ton of uranium thus requires the mining of 500 tons of raw material which must be processed. Using the more marginal sources of uranium means mining thousands of tons of material to obtain one ton of uranium. Much of this uranium can be obtained only by strip mining on a massive scale. The use of marginal ores, especially the shales and granites found in the east, is thus an option which may not be viable for a nation concerned with preserving its remaining areas of natural beauty in the later part of this century. The specter of uranium strip mining on the scale which is required in the mountainous areas where the resource exists should be incentive enough to find a suitable alternative.

Perhaps more pressing than the availability of uranium ore itself is the question of the expansion of the huge plants required to enrich the ore to a grade suitable for power reactors. This process is still a government monopoly. The enrichment plants take seven years to plan and build and require enormous quantities of electric power for their operations. Research is underway for new methods to enrich uranium (using gas centrifuges and laser beams) but these are not yet commercially proven. Current federal executive branch policy is for private enterprise to provide the capital needed to expand uranium enrichment capacity but the Congress has not approved this policy. And industry will not move forward to build the expensive plants until the ground rules and the market are more clearly defined. The uncertainty over expansion of capacity to enrich uranium may well limit the growth of atomic energy in the 1980's. And it is by no means assured that even a clear government policy will remove obstacles to expansion.

In spite of these problems, the uranium supply is probably adequate for a rapid expansion of the atomic power industry over the next decade and perhaps some years beyond even with the current inefficient plants. By the late 1980's we will begin to approach the limit of rapid expansion unless there are major unexpected new discoveries of resources or unless we develop technology that uses our uranium resource more efficiently.

Current nuclear reactors hopefully will some day be remembered as the "Model-T Fords" of the atomic energy era (the Model-T was a good and even revolutionary car in its time, remember). Development work is proceeding on reactors that can use thorium, another large source of atomic energy. But much more important are the highly efficient nuclear plants called breeder reactors.[122] They are called breeders because they enrich uranium into a nuclear fuel and thus tend to "breed" their own fuel.

The breeder is crucial because it will greatly enlarge the uranium resource base. First of all, it is 70 to 80 times as efficient as existing plants. Thus a pound of uranium in a breeder will generate 70 times more kilowatt hours. Just as important, the breeders will use the "wastes" from current plants and generate their own fuel. Breeders are not perpetual motion machines but they will make it unnecessary to mine significant volumes of uranium for a long time to come.[123] Also, the plutonium which is a by-product of present day reactors can be used as a fuel for the breeder. Thus it puts waste products to good use as a fuel. And since the breeder greatly reduces the volume of uranium required and the cost of fuel per kilowatt hour it becomes economical to use ores that contain uranium in very thin concentrations. If the uranium in sea water could be extracted the supply would amount to billions of tons, and the uranium fuel might last for thousands of years rather than a few decades.

Although breeders offer great hope for the future, they also bring all of the problems of current nuclear reactors previously discussed, and to some people in a more terrifying way. In any event the first commercial plants in the U.S. are at least fifteen years away. Their future availability for reliable, safe and economical supply is by no means certain. If for some reason the breeder were to be abandoned or to fail, nuclear energy could prove to be a very short-lived energy source—a nuclear flash in the dark so to speak. Without the breeder the contribution of nuclear fission power would be so small as not to be worth the risks to society. (See Chapter 4 for a discussion of environmental and safety concerns.)

Atomic energy is a commercial reality for generating electric power. It is supplying the energy for half of the new electric power plants now being built. But its contribution to the future of our high energy civilization remains overshadowed by a large question mark. If more efficient nuclear plants cannot be built, we could peak out of usable nuclear energy in the U.S. as fast, or faster than, the fossil fuels. In the meantime the industry will be severely taxed to meet the rapid rates of growth now projected. The public still has great fear of atomic power and the question of safety heavily influences the pace at which our peaceful use of the atom can be expanded.

WATER POWER—ONLY A TRICKLE

The crucial fact about hydroelectric power is that it supplies only 4 percent of the total U.S. energy supply today and the percentage will gradually decline in the future.[124]

Hydroelectric power has played an important and glamorous role in meeting the energy needs of many regions of our nation. But most of the economical and socially acceptable water power sites in the U.S. are already developed. There are no doubt water power projects yet to be built that are important for regional needs. But we are now aware that dams and the reservoirs they create can do irreparable damage to the wildlife in a stream and to the natural beauty of a wild river.[125]

Nevertheless, the remaining water power potential in this country is large. The Federal Power Commission estimates that 94,000 megawatts with annual energy generation of 325 billion kilowatt hours could be developed. If all these dams were built it might contribute substantially to our total energy needs in the coming decades and could meet a sizeable share of requirements in such regions as the Pacific Northwest. But most of the hydro potential is in the Columbia River Basin on tributaries where the opposition of sports fishermen, nature lovers, and the general public is strong and growing stronger.

Hydropower is a renewable resource which does not deplete.[126] But the land resources of the U.S. are now so limited and precious and the power potential of remaining projects so relatively small that hydropower cannot be considered a significant source of energy for future growth in the U.S.

THE WORLD ENERGY SUPPLY PICTURE

The U.S. has been importing most of its growth in energy supply in the past few years. It is therefore logical, indeed necessary, that we examine the worldwide supply-demand picture.

Oil

The prospects for oil are still bright in many parts of the world, much as they were in the U.S. early in this century. The capacity to produce barely exceeds worldwide demand but discovered reserves of oil in the Middle East are huge and the oil fields are cheap to produce. With few exceptions these fields are still in the early stages of productive life. In addition to large discovered reserves, much oil is known to exist which is yet to be drilled.

Even so, a picture of abundance can be as misleading for the world as it was for the U.S. just a few years ago. Many factors could limit the rate of expansion of the oil in the ground. Even the big Middle Eastern barrel has a bottom, and with most of the world drawing it down, it will peak out in a few decades and gradually run out.

For the immediate future, however, the world's discovered oil supplies seem comfortably large, estimated at 630 billion barrels, compared with the U.S.'s 45 billion barrels (see Table 4, below). Productive capacity is capable of rapid expansion in the Middle East. The big question is not the availability of oil but the pace at which the oil-rich nations will expand their capacity and sell it, if they expand it at all.[127]

The major producers in the Middle East are Iran, Saudi Arabia, Kuwait, Iraq, and Abu Dhabi. These are all Arab nations except for Iran. Most of the world's discovered reserves are found in the vast oil fields in Saudi Arabia, Iran, and Iraq. Saudi Arabia alone, before the Mid-East crisis, was expected to expand its oil production from 6.5 million barrels per day on January 1, 1973, to 11.5 million by the end of 1975. This 5-million barrel per day expansion represented over half the expected expansion in the entire world by 1975. And the Middle East as a whole was counted on to supply 75 percent of the world's total growth in the next few years. Adequate supplies in the world market hinge to a great extent on the rate of expansion in this volatile region in the decade ahead.

If we move a few hundred miles to the west, we find "big oil" in North Africa, although not the same volumes as in the Middle East. Here too the Arab nations of Libya and Algeria are the large producers, with Nigeria now also a major producer. African proven reserves are some 60 billion barrels, larger than those of the U.S., but still only one-seventh as large as the Middle East reserves.[129]

Table 4

WORLD PETROLEUM PRODUCTION AND PROVED RESERVES—1971[128]

(billions of barrels)

REGION	Production		Proved Reserves (Includes on-shore and off-shore)	
North America	4.8		60.2	
of which Canada		0.5		10.2
U.S.A		4.1		45.4
South America	1.7		27.0	
of which Venezuela		1.3		13.9
Africa	2.1		59.3	
of which Algeria		0.3		12.3
Libya		1.0		25.0
Nigeria		0.6		11.7
Europe	0.3		16.4	
Asia	9.2		467.6	
of which China		0.2		20.0
Indonesia		0.3		10.4
Iran		1.7		55.5
Iraq		0.6		36.0
Kuwait		1.2		78.2
Saudi Arabia		1.7		157.5
U.S.S.R.		2.7		75.0
United Arab Emirates		0.4		20.5
Oceania	0.1		3.1	
WORLD TOTAL	18.2		633.6	

In the past the U.S. has largely shut itself off from the

abundant low cost oil in the Middle East and Africa and other parts of the world. But in recent years our imports increased out of necessity. In 1973, before the Arab embargo, more than 10 percent of the U.S. oil supply originated in the Middle East and North Africa. The Middle East and Africa for years have supplied the bulk of Western European and Japanese oil requirements, primarily from Arab nations.

The Middle East is by far the major source, but oil is also being found in many other parts of the world. An exciting new source of supply is the North Sea, which is likely to fill most of Great Britain's needs in the 1980's. The full extent of the North Sea discoveries is not yet known, although they are small compared to the Middle East. The Middle East and North Africa will remain the principal sources of oil for Western Europe and Japan in the foreseeable future.

Other centers of oil production are Venezuela and Canada, which are still the major sources of U.S. imports.[130] Venezuela is a mature oil province which still ranks as a major producer accounting for 13 percent of the world total.[131] Its discovered reserves are relatively small (14 billion barrels) and its conventional oil provinces appear to be peaking out. However, Venezuela's heavy oil belt area contains several hundred billion barrels of heavier oil which is high in sulphur and more expensive to produce. It is in a category analogous to the extensive Canadian tar sands and U.S. shale oil. Some additional research and development are needed to bring it into the commercial mainstream. There is the added complication of Venezuelan government policy which favors development by the national petroleum company, further complicating the possibility of rapid increases in production.

Canada is currently a small oil producer but one with large prospects for the future. The discovered reserves in Canada are only some 10 billion barrels but there is good reason to believe that another 100 billion barrels are yet to be discovered, primarily in the frontier regions in Northern Canada. In addition to these conventional oil resources the tar sands in the province of Alberta contain over 300 billion barrels of crude oil, some 60 billion barrels of which appear to be economically recoverable at prevailing U.S. domestic prices.[132]

Canada thus contains very large oil resources far in excess of the foreseeable needs of its relatively small population. It also provides a secure and environmentally acceptable source for the U.S. market since transportation is overland and Canada is the largest trading partner of the United States. Canada finds a natural market for its surplus oil in the U.S., but Canadians are not at all eager to become the powerhouse for their giant neighbor's industry. Canada is concerned about its own environment and also wants reverse trade that will enable oil production to play a role in making possible a diversified economy for both nations. Thus the future availability of Canadian oil in the U.S. market depends on resolving broader questions of trade policy between the two nations.

The Soviet Union has large oil resources within its borders but it also has large needs for oil. Russia is still self-sufficient and will undoubtedly remain so. But it has little oil to export, and it needs expanding supplies for its Eastern European dependencies. China is beginning to develop its oil resources but its production and consumption are still quite small. The Communist bloc nations (primarily the U.S.S.R.) are believed to possess oil reserves of some 100 billion barrels, twice those of the U.S., while total Communist consumption of oil is only about one-half U.S. consumption. The undiscovered oil resources in the Soviet Union, China, and the very promising East China Sea are conceded by most authorities to be much larger than our own.

For the next decade the consuming world will be dependent on the geologic provinces in which oil has already been discovered. The lead time for finding and bringing a brand new area into sizable production is a decade. But there are large quantities of oil yet to be discovered on this planet. Some experts have been brave enough to estimate the ultimate oil resources in the world, and their estimates range from 1,350 billion barrels to 2,100 billion barrels—even to

6,200 to 10,000 billion barrels if marginal resources are included.[133]

We must keep in mind that in the past only about 30 percent of the oil in a reservoir has been economically recoverable. If that trend is not changed the potential resource is, of course, only one-third of these large numbers. In any event the estimates are most speculative, being mainly extrapolations of past experience. But they do reflect a large and growing body of geologic information.

The global oil resource estimates (roughly six times the size of U.S. resources[134]) suggest that the world will not run out of oil for many decades to come. Yet these numbers can provide a very false sense of comfort. If current growth trends persist, world oil consumption will total 1,000 billion barrels in the next 30 years and more than 3,000 billion barrels in the 30 years to follow. Just as with the domestic situation, the critical question is: When will world production reach its peak? The same constraints of available capital, environmental concerns, and lead-times will inhibit growth. But more crucial on the world scene will be the policy of the Middle Eastern oil-producing nations. (See Chapter 9.)

Worldwide demand for oil promises to continue growing exponentially at some 5 percent annually. If so, the discovered petroleum resources may reach their peak rate of production in the 1980's even if all the constraints give way. Surely additional oil will be found but the resource base itself suggests that oil production could well reach its peak before the turn of the century. The world oil picture is thus quite likely to follow the U.S. pattern except that production may grow at least into the 1980's and perhaps a decade longer before it, too, reaches a peak and begins to decline.

The U.S. can purchase oil from other nations to help meet its needs for the next decade and perhaps some years beyond. But the world will steadily move from surplus to shortage as standards of living approach current levels in the U.S. And this will greatly complicate our efforts to sustain past energy growth rates from outside sources.

We have discussed oil as a source of energy but perhaps

our most fundamental mistake is to ignore the future value of oil for other basic uses. Oil, of course, is the basic raw material for the petro-chemical industry. Oil also can be converted to protein and used as food. It may well be that future generations will consider us barbarians for burning in a few short decades most of these unique collections of hydrocarbons that required 100 million years to produce. Thus a more fundamental motive than preserving energy supply should spur efforts to develop alternatives to the finite supply of oil in the world. We may find infinitely better and more rewarding uses for oil than burning it.

Gas Supplies From Abroad

As we look abroad for sources of natural gas, Canada is clearly our most important potential source. The existing proved or discovered Canadian reserves are some 60 trillion cubic feet. The ultimate potential of the Canadian gas resource has been reliably estimated to be some 650 trillion cubic feet.[135] Canada thus has large natural gas resources beyond the needs of its relatively small population, which it may wish to sell in the U.S. markets. And Canadian gas can be transported to the U.S. economically and safely by pipeline. In fact many consuming centers of the Pacific Northwest and the Midwest are closer to Canadian sources than to the U.S. gas fields. The crucial factors are the policies of Canada and the United States with respect to trade.

If we look to overseas sources we find natural gas located generally in proximity to oil. Various estimates of the potential supply of natural gas are available based primarily on a ratio of 6,000 to 7,500 cubic feet of gas for every barrel of oil. These speculative estimates range from 8,000 to 12,000 trillion cubic feet. These are large numbers indeed, representing the energy equivalent of 1,360 to 2,000 billion barrels of oil.

The world's natural gas resources are probably a source of energy as large as the lower range of estimates of the world's oil supply. Sadly, in many parts of the world this resource is being burned at wellhead and wasted for lack of a market.

Until very recently natural gas could not be shipped overseas for lack of technology. And there was no local consumer market in much of the Middle East and other major producing countries.

Technology is now available to freeze the natural gas and transport it across the oceans. But such transportation is quite expensive compared with oil, for which transportation adds only one-fourth to its wellhead price even from the Persian Gulf to New York. Natural gas, however, requires large investments in liquification and regasification plants at each end of the journey and special cryogenic tankers which are giant floating thermos bottles. Transportation costs are 5 or 6 times the wellhead price.[136] Each project requires large special investments of its own and involves the complications of dealing with foreign governments and winning U.S. regulatory approval.[137]

In the final analysis the natural gas resources in the Middle East and elsewhere in the Eastern Hemisphere can be used much more economically in the producing nations as raw material for petro-chemical production and in consuming markets in Europe where transportation costs would be somewhat lower than to the U.S. Therefore, the United States cannot expect to draw very heavily on Eastern Hemisphere natural gas supplies. Imports of natural gas in quantities sufficient to supply the bulk of U.S. energy expansion needs do not seem probable, even though they may provide a significant supplement.

The impact of the natural gas shortage is to stimulate a greater demand for oil, which is the most readily available substitute. Here we see the domino effect in action because the gas shortage will exacerbate oil supply problems and environmental concerns.

Coal

The world's coal resources, oddly enough, seem to favor the big nuclear powers—the Soviet Union, China and the United States. The coal resources of Russia and China are more than twice the size of our own and the three nations combined contain some 90 percent of the 8,600 trillion tons of total recoverable coal resources.[138] This is a storehouse of energy roughly sixty times the size of all the world's proven reserves of oil and gas combined.

As we have seen, Western Europe is shifting away from coal, having already mined most of its economical supply at or near current prices. Russian coal production is somewhat larger than in the U.S. but it is not growing rapidly. In Russia, too, petroleum is supplying most of the energy growth. But coal provides the bulk of Chinese commercial energy, and production now approaches 400 million tons a year, two-thirds as large as U.S. production. The potential for growth of coal production and use in China is truly staggering.[139] Indeed, there is a serious question of the impact on the global environment if the industrialization of China is fueled by coal without removing the sulphur and other sources of atmospheric contamination.

In thinking of what Russia, China, and the U.S. have in common, coal is not likely to be high on any diplomat's list. But the future of our high-energy civilization could perhaps hinge on the combined efforts of these nuclear superpowers to learn how to use this old-fashioned form of energy in harmony with nature.

Worldwide Uranium Supplies

The worldwide picture in uranium supply is not essentially different from that in the U.S.—a current surplus but the prospect of shortages looming over the horizon. Shortages will occur later than in the U.S., but even on a worldwide basis uranium supply at acceptable costs to society may reach a peak of production in a few decades unless the breeder reactor proves successful.

The major known uranium producing countries outside the United States are Canada and South Africa, each of which possesses reasonably assured discovered reserves roughly comparable in size to the U.S. reserves. The three countries together possess 80 percent of the total outside the Communist nations. There is little information concerning uranium resources in China, the Soviet Union and Eastern Europe. It is believed, however, that major uranium deposits are located in the Soviet Union.

Construction of atomic power plants in Western Europe has proceeded at a slower pace than expected and as a consequence the world demand for uranium is below productive capacity. Known reserves at reasonable prices appear sufficient to meet needs amply into the 1980's. But if nuclear power takes off—as indeed something must as oil and gas begin to taper off—then the adequacy of our uranium resources to support new atomic power plants in the 1990's becomes questionable.

With respect to worldwide reserves, the available figures are highly speculative. However, growth in world demand for uranium is expected to be quite rapid—much the same as in the U.S. The shape of the curve is similar but the peak occurs about a decade later. The worldwide crunch in uranium supplies without the breeder comes in the 1990's rather than the 1980's.

In view of the present feast and looming famine in uranium supplies, it is interesting that the U.S. government has placed a complete embargo on the importation of foreign uranium ever since 1966.[140] The restriction on imports is of course to protect the markets of American producers. But we do so at the ultimate expense of the American consumer. A more fundamental consequence of our embargo against imports of uranium is that we hasten the day when our own resources will be depleted. By the time we need to import foreign uranium, in the 1980's, demand in the rest of the world may have grown to the point that there will be little available for import.

Hydropower

Hydropower is a major source of future energy in South America and Africa, each of which contain over 500,000 megawatts of capacity—far more capacity than in all of the U.S. electrical systems today. Almost all of this capacity remains to be developed. And there is also large hydropower potential to be developed in Southeast Asia.

The total energy potential of the world's hydro-capacity is significant. If it could be developed in the next 30 years—a gigantic undertaking in terms of capital investment—it could supply perhaps 25 percent of the total energy supply, if living standards around the world continue to improve.

SUMMARY

Since 1798 when Malthus first clearly recognized the possibility, there have been many predictions of a shortage of fuels and other resources. Thus far all such predictions have been wrong. Yet it is a certainty that if growth in consumption continues and new sources are not developed, someday such predictions will be accurate. The demands of growth that doubles and redoubles current energy consumption will drain the planet's resource basin to such an extent that prior consumption will be only a trickle by comparison.

Yet the danger to consumers is much more real and immediate. We have fashioned a high-energy civilization that must be fed by an ever increasing flow of energy. But our society is of more than one mind and it is changing. We are not prepared to pay the price, incur the damage to the environment, or otherwise make the all-out commitment to energy production necessary to sustain our levels of consumption. In this state of indecision and of transition to a society with somewhat different goals, we are not expanding the flow of energy fast enough to meet projected needs.

Perhaps with a single-minded national commitment to increasing production from existing sources of supply, we could satisfy the demands of rapid growth for a decade or two longer. But the new values of our society are slowing down the energy producers and growth rates will reflect that conflict. In any event our existing energy sources will peak-out long before they run out.

The obvious answer is to develop new energy sources. As we shall see in Chapter 12, the potential new sources are as obvious as the sun and as mysterious as the power of the hydrogen bomb, both of which scientists believe can be converted into electricity to continue our high energy civilization indefinitely. In a more limited way the oil shale deposits in the Rockies, the heat in the earth, and the promise of technologies to remove environmental constraints on the use of coal may buy us time to develop these more abundant sources.

The world in which we live is utterly dependent on the development of new technology for energy. The U.S. government has spent billions racing to the moon, and has even developed a solar energy machine to power the flight and an electric car to explore the lunar surface. Meanwhile, we have sadly neglected the research and development programs needed to provide clean energy here on our spaceship earth.

CHAPTER 6

Foreign Policy and Energy Policy

The Arab oil boycott in the fall of 1973 and the massive price increases that followed not only surprised the U.S. but also punished us for naively supposing that we could build a wall between energy policy and foreign policy in an increasingly dependent world. Quite obviously, the economic well-being of the nation and indeed issues of peace and war hinge on our ability to develop a realistic energy policy for the U.S.—one which is attuned to the world at large.

In recent years, the U.S. has imported a steadily increasing proportion of its total energy supply, which reached a peak of 15 percent prior to the Arab oil cutoff in October, 1973. Perhaps more significant, more than half of U.S. new supplies of fuel for the future were to be purchased from other nations. This expanding U.S. need for oil and natural gas imports can contribute to world peace and prosperity, but, as events have proved, it can also make this nation vulnerable to international blackmail and precipitate a worldwide energy crisis of severe proportions.

Our relations with the Soviet Union provide a good example of the way that prospective imports of natural gas and oil can contribute to world peace. Trade is the symbol of U.S.-Soviet efforts to bury the Cold War. The USSR needs our grain, capital and technology, and we need their oil, so that trade may be the solvent of national differences. But if the U.S. must import fuel out of necessity rather than by choice, and if the oil producing nations really do not need our money, recent events prove that our dependence gives them a powerful lever with which they can exact higher prices and exert powerful pressure for changes in U.S. foreign policy.

The Arab oil embargo revealed just how vulnerable the United States, Western Europe and Japan had become. We were all relying complacently on an ever-expanding supply from one of the most unstable regions in the world. The more radical Arab nations had been warning us for years that oil and politics would be mixed. The more conservative Arab nations on the Persian Gulf had stated that the gush of oil might be slowed down for economic reasons. But no one in authority in the U.S. was listening. And no one really expected that the conservative kingdom of Saudi Arabia, on which we had been relying for most of the growth in production, would use the "oil weapon" in an attempt to influence U.S. policy toward Israel.

There are at least two major lessons from the Arab boycott and the subsequent doubling of crude oil prices that we dare not ignore. One is that Arab oil is subject to interruption without notice. It can be cut off at any time. It should be classed as interruptible power—just like hydroelectric power which is available only in years of ample rainfall—and it should be treated accordingly in our energy supply strategy.

The second lesson is not to judge the producing nations too harshly for increasing prices. For decades the Western world bought Middle East oil at bargain prices in comparison to alternative sources. The economies of Europe and Japan flourished in part because of the availability of low-priced Middle Eastern oil. The turn-around has indeed been abrupt and the economies of oil importing nations as diverse as India and Japan have received quite a shock. Adjustments are necessary to avert a worldwide recession. Yet the billions of dollars now flowing into the Middle East could be used to improve the lives of millions of Arabs who live in poverty in the midst of oil riches. The key question is whether the huge funds wreck the world's economy, finance a gigantic arms race or by some miracle of diplomacy can be channeled into regional development programs.

If the U.S. is to play a leading role as a peacemaker it must adopt energy policies that will give it independent authority to do so. This means participating in world energy trade and helping to solve the energy problems of all nations. But it also means not becoming dependent for energy on any single nation or a group of nations with a common desire to restrict American freedom of action at home or abroad.

The need to import a growing proportion of our fuel supply is a new development for America. It reminds us of wars lost by nations without reliable sources of petroleum and wars started to protect petroleum lifelines. Anyone old enough to recall World War II knows that the Allied blockade which cut off enemy imports of fuel played a major role in the defeat of Germany and Japan. The Japanese were so desperate for oil in the latter days of the war that they actually offered the Soviet Union several of their warships in exchange for fuel to power their fleet.[141] When the Germans were cut off from Silesian coal, which they converted to synthetic oil, Albert Speer, the German armament chief, concluded: "The war is lost."[142]

In 1956, the British and French invaded Egypt largely to protect their petroleum lifeline. Their intervention was precipitated by Colonel Nasser's takeover of the Suez Canal, then a vital link in the oil supplies of Western Europe. Larger tankers have since rendered the Canal less important, but in 1956 its control seemed important enough to justify war.

The 1973 Yom Kippur war in the Middle East and the Arab cut-off of oil to influence the terms of peace again places oil in a military and strategic context. But that juxtaposition is quite misleading. It would be a grave mistake to view the international aspects of energy policy from a narrowly military perspective. The critical new fact is that since

World War II, demand for energy in the U.S. civilian economy has grown enormously as compared to our military needs. Even in wartime, military requirements for energy would constitute no more than 10 percent of total U.S. consumption.[143] Emergency military needs for energy could easily be satisfied from domestic sources. In many ways, if there were a stockpile to afford time for readjustment, the task of balancing our energy accounts in a war-time emergency would be easier than in peacetime. We could burn domestic coal (too dirty to use in peacetime) and quickly put a stop to the wasteful energy joyride. Imports of energy are not primarily a wartime problem, especially in a world haunted by fears of a short and terrible atomic war. As we now are painfully aware it is the peacetime economy that is so utterly dependent on reliable supplies of energy. The oil weapon is economic in nature but in many ways more powerful than the Bomb. The major oil producing nations can literally bring the Western world to a halt if they dare to do so. Even a short and fluctuating shutting off of the oil flow can create a world-wide recession and bring powerful nations to their knees begging for oil.

In 1938, just prior to the outbreak of World War II, the U.S. was self-sufficient in energy supplies. The U.S. then accounted for more than 60 percent of the world's oil production and was a net exporter of energy. The postwar economic boom created a demand for energy that soon outpaced domestic production, and by 1950 the U.S. had become a net importer. To be sure, the percentage of imports was small at first. By 1965, however, imports accounted for 22 percent of U.S. oil consumption and 7-1/2 percent of our total energy supply. During the 1960's the U.S. chose, primarily for domestic political reasons, to limit the imports of much lower priced oil from the Middle East and elsewhere and to rely mainly on domestic sources. During those years there was surplus producing capacity in the U.S., although perhaps not as much as we thought. There was certainly no question about adequacy of fuel supply. So the nation opted for a policy of "Drain America First."

The nation now faces a future in which the U.S. planned to increase consumption of energy at an alarming fast pace while domestic production of petroleum has virtually peaked out. The projections of energy requirements for the 1980's are fraught with uncertainty, as we saw in Chapter 3. But if we continue with a historic growth in consumption of energy for the remainder of the 1970's we must import sizable quantities of oil (and natural gas) from the Middle East or suffer rather severe shortages of energy.

The growth in energy consumption in the U.S. in 1972 over the prior year was the equivalent of more than one-half billion barrels of oil; about half of the increase was in the form of imported oil. More than 70 percent of the oil imported in 1972 originated in Canada and Latin America. Imports from North Africa and the Middle East in 1972 were small but imports from the Middle East grew rapidly in 1973. Yet at the time the embargo took effect the U.S. was importing directly only about one million barrels per day of Arab oil and another one million barrels per day in oil products that originated in Arab nations. This amounted to about 14 percent of U.S. oil supply and about 7 percent of total energy. Through 1973, the United States was still obtaining only a small part of its total energy supply from the Middle East, but the share was growing. In contrast Japan imports almost all of its energy, and 60 percent of the total supply is oil from the Middle East. Western Europe also imports more than 50 percent of its total energy supply from the prolific oil fields around the Persian Gulf and in North Africa.

But we should not delude ourselves. The current sources of oil production in the United States, Canada and Venezuela are all operating near the peak of their productive capacity. If America continues its current pattern of energy consumption, our dependence on Middle Eastern oil will grow and we will be competing with Western Europe, Japan and the developing nations for any new production that may become available.

According to the National Petroleum Council,[144] more than half of the growth in energy consumption projected for the U.S. in the next decade will probably be supplied by imports. This is a most significant possibility. Domestic natural gas has supplied most of the energy growth in the U.S. since World War II. In the decade ahead the U.S. must either tap the world oil market or trim its vast demand for additional energy.

The National Petroleum Council also projected that with a continuation of current trends and policies, more than half of the U.S. oil supply and 30 percent of total energy—some 15 million barrels of oil per day—would have to be imported by 1985, mostly from the Middle East. When first published in 1972, these estimates were considered alarmist, but the actual pattern of supply soon rushed in that direction. Before the Arab boycott we were importing one-third of our oil supplies; if we return to a business-as-usual posture, we could well be importing one-half of our oil supplies by 1980. If this scenario is actually permitted to unfold, the quantities of energy destined for the U.S. from the world market (which means primarily the Middle East) in the next decade would greatly exceed the purchases of any other nation.

The consequences of the U.S. going to the world market for most of its additional energy are now too obvious and disturbing. Putting aside for a moment the Arab-Israel conflict, the immense projected purchases by the U.S. would threaten the supply of every other consuming nation on earth large or small. The Middle Eastern barrel may be a big one but Western Europe and Japan are properly concerned to see energy-guzzling America starting to draw from the same wells on which they depend so heavily. Developing nations are horrified at the prospect. They are already being priced out of the energy market and their development slowed or even halted as the industrialized nations bid up oil prices in a scramble to keep their cars and machines running.

The immediate foreign policy questions center on the oil-producing nations, and all the consuming nations have a stake in their response. Except for the Soviet Union, most nations that consume appreciable quantities of energy must import a sizable portion of their supplies. In fact, oil is the largest single item of world trade. It is the life-blood of industrial societies, so that any major stoppage in the world-wide flow of energy is a serious threat to peace. Our foreign policy must therefore be concerned with assuring the free flow of energy not only to the U.S. but to all nations.

Trade in energy, like other commercial transactions among nations, has helped poor countries to develop and prosper. Producing nations such as Venezuela, Iran and Indonesia are outstanding examples of the use of oil revenues to fund economic development. In fact Venezuela's ability to maintain a democratic form of government could well depend on the continued flow of revenues from oil exports. These producing nations with sizable populations which are embarked on ambitious development programs do not want to stop the flow of petroleum and thereby give up the large revenues at stake. The conduct of the Arab nation of Iraq during the 1973-74 embargo is instructive in this respect. Despite its militant opposition to Israel, Iraq did not cut back its sales of oil. It claimed that it embargoed shipments to the U.S. but publicly refused to join in any cutbacks of production and sales. Other, non-Arab nations such as Iran, Venezuela, Canada and Nigeria did not join in the cutback or the embargo.

The Persian Gulf nations that control most of the world's oil may well have sound reasons for limiting the rate of growth in production. Before the double shock of embargo and soaring oil prices the world was counting on Saudi Arabia, Iran, and Abu Dhabi for an expansion of production of roughly 21 million barrels per day between 1972 and 1980, which is 70 percent of the expected increased production in the world during those years. The world will, of financial necessity, now make do with far less oil, but it is by no means certain that expansion will occur in the quantities necessary to avoid a continuing shortage. Indeed the immense revenues generated by the lower volume of sales make it very attractive for the major producing nations to keep the world on a short energy ration.

Saudi Arabia and Abu Dhabi are nations with small populations—8 million and 700,000, respectively—and their internal development plans are just getting underway. Saudi Arabia alone was being counted on to increase its production three-fold—to 20 million barrels per day by 1980. The sums of money that would be generated by such growth—some $50 billion annually by 1980—are well beyond Saudi Arabia's ability to spend internally. It may even be more than the Saudis wish to have available for investments in other nations.

The oil in the ground may seem an easier and safer investment to manage than foreign currency of uncertain value or securities in a distant land.

The overriding concern of such Arab nations as Libya, Saudi Arabia, Kuwait, and others is that their oil resource may be depleted before their countries can be developed. Their leaders know it will take decades to build up industrial economies in nations. They know their oil is a wasting asset with a limited life. Rather than producing it at a pace much faster than their ability to use the revenues, they may opt to increase oil production at a pace attuned to the ability of nomads to move into the twentieth century. They have little incentive to try to satisfy the appetite of energy-wasting Americans. Indeed, Libya, Algeria, and Kuwait had already placed permanent ceilings on their oil production before the boycott and price escalations of the winter of 1973-74. If Saudi Arabia and Abu Dhabi follow suit, the world may face a severe long-term energy crisis.

There are, however, major producing nations with the ability to expand production that are in an altogether different posture. Iran has a large population (30 million), major development plans, and ambitions to build up its military forces and exert regional influence. Iran is committed to the most rapid possible expansion of its oil reserves because it needs the money. Iran's planned expansion from about 5 million to 8 million barrels per day in 1980 will almost surely occur but that may be the limit which Iran's reserves can sustain. Iran has stated that it will impose a production plateau at the 8-10-million-barrels-per-day level. In the 1980's the country's internal demands for oil as a petrochemical feedstock as well as a fuel will grow, and its exports may well begin to decline.

Iraq is the only other nation with large discovered reserves of oil that could make a difference in this decade. Oil production in Iraq could be increased several million barrels per day in a few years. And Iraq, a nation of 7 million poor people, could certainly use the revenue. It is already producing over 2 million barrels per day and is committed to rapid growth. But Iraq has been torn by internal political dissension and its oil production hampered by disputes between the government and the international oil companies. Iraq has nationalized some of its oil fields, and the companies have slowed the flow of Iraqi oil to the Western world.

Iraq is friendly to the Soviet Union and a rival for supremacy in the Persian Gulf of both Iran and Saudi Arabia, nations which the U.S. is supporting through the sale of arms. But the U.S. has managed to become friendly with both China and the Soviet Union, and in concert with Western Europe and Japan we should be able to find a means to facilitate the development and flow of Iraqi oil into the world market without impairing our relations with the superpowers of oil supply. The French and Japanese are actively pursuing this possibility. The U.S. government should realize that the interests of oil consumers in adequate supply may be quite different from the financial stake of the big companies which own the oil. Iraqi oil could relieve much of the supply pressure in the world market in the next few years.

As we learned in the previous chapter, there are no discovered oil reserves outside the Middle East available in quantities to fill the gap for the remainder of this decade. The North Sea oil will begin to flow in quantity in the 1980's, and the same can be said for any other major sources. Some increased production from other areas will no doubt occur, but the Middle East in general and Saudi Arabia in particular will be critical to balancing the world's energy budget for the foreseeable future.

This brief survey of the world scene suggests that while there is a great deal of oil in the ground, it is located in nations that may want to conserve it rather than exchange it for more dollars for investment. This is especially true since the producing nations have learned in recent years that by keeping the consuming nations on a short ration they enhance their bargaining power and obtain higher prices for what they do sell.

The actions of the Arab nations that sent the price of their oil soaring underscore one of the alluring facts about Middle Eastern oil—its cost is much lower than oil in other regions or other sources of energy. The fact that people in the consuming nations are now willing to pay billions of dollars more for that oil is a good measure of the benefits they reaped from the lower prices that were in effect for decades. For years the international oil companies made large profits on the Middle East oil they produced. They bought low and sold high in the American market where prices were set by the costs of domestic oil, which were more than $1 a barrel higher, even including the cost of transportation. And in Europe and Japan, where most of the Middle Eastern oil was sold, the low-price oil contributed substantially to the remarkable economic growth that has taken place since World War II.

The producing nations rightly felt cheated. For decades their only resource was exploited and they received only a pittance—as little as 5 cents a barrel in the early days. The bargaining unit called the Organization of Petroleum Exporting Countries (OPEC) was formed in 1960, and it has gradually gained strength. Libya's success in 1970 in limiting production at a time of tight crude supply in the world market marked the turning point from a buyer's to a seller's market. Demands for higher prices were met because consumers feared a shortage if Libya further curtailed production. Saudi Arabia and Iran then became determined to match Libya's success at the bargaining table, with the result that the producers have now completely turned the tables on the consumers. But as we shall see, the oil companies are still managing to do quite well at the expense of the American consumer.

As recently as 1970 the Arab nations received only about 3 cents per gallon of the 30 cents the American consumer paid for gasoline at the pump. Today the producing nations are netting about 17 cents out of the 50 cents per gallon that Americans now pay. The producing nations are charging the consuming nations the approximate cost of alternative sources. In a sense this is the test of value that any businessman with a lock on the market could be expected to exact. But sudden change is painful, and it is especially painful to have to give up an economic advantage to which one has become accustomed.

The mounting financial success of the OPEC nations in bargaining with the international oil companies in recent years raised the fear that they might use their bargaining power for other purposes, especially to influence U.S. policy toward Israel. But these fears were taken rather lightly until October 1973, when they materialized.

A concerted Arab boycott turned out to be no idle gesture. The Arab nations were producing some 20 million barrels of oil a day at the time the boycott was imposed on October 17, 1973. This amount represented more than 40 percent of the non-Communist supply. Although their maximum cutback reduced overall oil supply by only 10 percent, it was enough to send shock waves through the consumer nations. It caused almost every industrialized nation in the world except the United States and the Dutch to tilt heavily toward the Arab position in the Middle East. Quite apart from concern over our own energy supply, the U.S. could not ignore the effect the oil cutback had on its allies. The Arabs have demonstrated the ability to use their oil weapon with skill against Western Europe and Japan.

The oil weapon was brandished when it was not yet a big club against the U.S. Of course, if the U.S. reverts to rapid growth in oil consumption from the Arab nations, the club will grow. The erosion of U.S. independence of action in the Middle East could occur in a much more subtle and less dramatic fashion than as the result of another sudden cutoff of oil. Because it happened once everyone will fear that it might happen again. Thus if the U.S. allows itself to become steadily more dependent on Arab oil, the threat of interruption will be a growing threat, exerting leverage on U.S.

foreign policy every day. As we have seen, the Arab nations have economic reasons for limiting production and keeping supplies tight. The next time they could send America their message about Israel in the name of conservation of their greatest resource.

If U.S. dependence on Arab supplies increases, commercial interests and consumers would automatically have a new incentive not to "upset" the Arab nations. This interest would inevitably cause gradual shifts and weaken popular support of Israel, a shift which may already be taking place. An American president needs all the flexibility and freedom of action he can command if the U.S. is to help keep the peace in the Middle East. Our policy should be both pro-Arab and pro-Israeli in the sense of providing economic and technical support to both sides. But the settlement of outstanding differences must in the final analysis take place between Israel and its Arab neighbors. The U.S.'s ability to help promote a peaceful accommodation could be severely limited if we become dependent on Arab oil, unless our dependence is part of a package that makes oil revenues as important to the Arabs as oil is to us.

This brings us to the opportunity that oil revenues provide for contributing to lasting peace in the Middle East through a development program that in turn would provide the strongest incentive for assuring a reliable flow of oil to consuming nations throughout the world. That option, simply stated, is for the U.S. in concert with other technically sophisticated nations to join in an all-out effort to help the Arab nations fashion a better life for themselves. It must be *their* plan of development, and it must be carried out in *their* style. But Western technology, managerial talent and know-how, and markets for industrial production are essential.

An effort with the vision and scale of the Marshall Plan is needed, but this time it would be financed with their money. The money is certainly available. The Arab nations are now receiving some $50 billion annually in oil revenues. What is needed is a grand design for transforming those billions into water, food, schools, hospitals, housing and industrial development with the same zeal and enthusiasm as we rebuilt

Europe and sent a man to the moon.

The obstacle to be overcome—and it is a big one—is to interest and enlist the leaders in all the nations concerned to make an ambitious commitment to such an effort. First there is the question of the American, European and Japanese commitment. Perhaps more fundamental is whether the rulers of the key Arab states are prepared to undertake programs that will create well-educated, affluent middle-class consumers and modern economies in their nations. The hope lies in the younger members of the ruling class who sense that pressure from the masses will continue to mount unless the oil wealth is used on a large scale for the benefit of all of their people. The other—and perhaps more difficult—problem is to interest the oil-rich producing nations in sharing their wealth with other Arab nations in regional development projects.

Pressures are now rising in many of the Arab oil producing states to initiate development programs. The progress being made in nearby Iran under its five-year, $23-billion development plan has shown what can be done. Moreover many of the producing countries now have a substantial cadre of technicians and administrators, educated abroad, who see more clearly than the older generation what should be done to use the large oil revenues for development.

There is of course a crucial difficulty. Israel is not the only nation in the Middle East without much oil. The large Arab nations such as Egypt and Syria have no oil either. Individual national development programs must therefore be supplemented by regional efforts so that the Palestinian refugees and the millions of Arab and Moslem peoples in non-oil producing nations will not continue to live in poverty while their more fortunate neighbors enjoy progress and prosperity. The agitation for war against Israel has its seeds in the poverty of these people. Peace will not come to the Middle East until all the peoples living there begin to enjoy a better life.

The time is ripe for initiatives from the U.S. and other consuming countries. The OPEC nations are themselves considering a Development Bank to receive funds, make long-term loans and mobilize technical talent. Hopefully the richer oil producing nations would make large contributions to the Bank. A U.S. government initiative to offer technology and technical and managerial assistance could accelerate the formation of such a Bank or other entity. The Bank's efforts in the main would supplement national programs to ensure that the Arab refugees and the millions of Arab and Moslem peoples in non-oil producing nations in the Middle East and perhaps elsewhere would benefit from development plans. Technical assistance could also be made available to those producing countries wishing to undertake development plans of their own.

Another way in which the consuming countries could encourage the use of oil revenues in development would be to open their markets to other exports from these countries. Because energy is so cheap in the Middle East, it would make sense for the producing countries to develop the manufacture of energy-intensive goods for sale in the developed countries where energy is more costly. This would have the double advantage of creating jobs and prosperity in the poorer countries, and benefitting the consumers in the richer countries.

The problems inherent in the massive flow of Arab oil to consuming nations should not blind us to the opportunities for facilitating development in producing nations. Here lies the truest security, regardless of the producing nation's ideology. If a nation is using the revenues from oil to increase the health, education, housing and material well-being of its people then no government would dare shut off the oil which pays for internal progress. The alternative is to let things drift as we have, with the added danger of large oil revenues being used to finance an arms race, thus escalating hostilities in the area and putting the world's oil supplies even furthur into jeopardy. It is not possible for the U.S. to wash its hands of this issue by curtailing our own consumption of Arab oil. If we are to play a role as peacemaker in the world, we must help all people to gain a standard of living that makes peace preferable to war.

The U.S. has much to gain by trading in oil with other nations so long as no nation or group of nations with a common political interest supplies a significant share of the U.S. market. Thus imports of natural gas and oil from the Soviet Union, if limited to about 5 percent of the U.S. market, would pose small risk while offering potentially large benefits in promoting *détente* and helping to balance our energy budget. The Soviet need for hard currency and their desire to buy wheat, high technology goods and other items from the U.S. offers reasonable assurance that the energy would not be cut off. Siberian oil and natural gas is one of the few available Soviet exports that America needs. It is a means of encouraging peaceful, long-term commerce between the super-powers.

Canada, Venezuela and the other producing nations in the Western Hemisphere offer another source of diversity in our oil imports. There are large energy resources in Canada—the tar sands and oil and gas in the north—which the Canadians may choose to develop for export to the U.S. But Americans must understand that Canadians view such development as a very mixed blessing. There are objections from Canadian environmentalists. There is the strong resentment of American domination of Canadian industry. There is the fear of inflation of the Canadian dollar arising from the inflow of too much capital for energy projects, an inflation that would hurt Canadian manufacturers attempting to compete in the U.S. market. And there is the belief that the fuel in the ground may be needed in Canada decades hence, and that its conservation may be a better choice for Canada than continuing to quick-march to the beat of the American drum. American policymakers need to be much more sensitive to the rising mood of nationalism in Canada. Otherwise the development of Canadian resources for export to the U.S. is not likely to be expanded very much.

The availability of fuels from Canada and Latin American nations in the decades ahead could be very important. Every barrel from a friendly neighbor would be one less barrel from the uncertain Middle East. Canada and Venezuela certainly have no interest in influencing our foreign policy in the same way as the Arab nations, and they have always been reliable sources in peacetime as well as during emergencies. To be sure Venezuela and other South American countries have taken steps to assert national ownership and to obtain as much money for their oil as the market would bear, a lesson in economics they no doubt learned from their American counterparts. But these financial actions should not be confused with threats to security of supply. Supplies have never been interrupted by these Latin American producing nations.

There is a definite foreign policy advantage to the U.S. in expanded trade with Latin America. Oil revenues, as we have pointed out, are essential to the prosperity of Venezuela, and they could be a catalyst for increasing the standard of living in Peru, Ecuador, Brazil, and other Latin American countries believed to contain petroleum resources. Unlike Canada, these nations need all the oil revenue they can obtain to fund development programs. Our energy policy might well provide greater incentives to the petroleum industry to concentrate their drilling in these nearby developing lands. The result may be to help resolve the "energy crisis" and at the same time provide revenues to nations now in poverty that could avert the danger of more "Cubas."

Flexibility and diversity should be the hallmarks of U.S. import policy. Imports are essential to meet our needs and they can be used to enhance our foreign policy objectives, if we fit our imports into an overall strategy for security of supply.

The problems associated with importing energy do not end with the question of reliability of supply and potential leverage on our foreign policy. Recent speculation on the economic consequences of massive imports may be somewhat exaggerated but they are nonetheless formidable, especially in view of continuing price increases. The U.S. will be paying out some $15 to $20 billion in 1974 for oil imports. These sums are at least $10 billion more than we had planned, and they will have quite a negative impact on our balance of payments. The sums would of course increase if the volume of imports increases. However, focusing only on one item of trade can be grossly misleading.

The dollars paid to oil producing states are used to buy goods or services from abroad, very often from the U.S. It may come as a surprise that in 1972 the U.S. sold the Arab nations over $500 million more in goods and services than the cost of the oil we bought from them. U.S. exports to the Arab states ran almost twice as high as imports, despite the record level of oil imports. True, our imports will grow tremendously in the future. But it is worth noting that for every dollar the U.S. spends on importing oil, Western Europe and Japan spend about $2. If U.S. exports of food, technical services, consumer goods, and industrial equipment continue to be competitive in the Middle East, we stand a good chance of getting most of our dollars back, and maybe even more. In addition to buying goods and services from abroad, the oil rich countries also invest abroad. Investment in U.S. securities benefits our balance of payments just as much as exports. The U.S. can certainly compete with any nation in investment opportunities. Thus to the extent our dollars do not return in payment for exports, they may return as portfolio and direct investments.

With oil revenues of the producing nations totalling as much as $100 billion a year by 1980, some people have expressed fears of another nature—that *too much* money would come back, enabling unfriendly nations to acquire control of some U.S. industries. This fear is far-fetched. Even if anticipated oil revenues are that large, they will be small compared with the much greater investment opportunities in 1980. The annual capital formation of industrialized countries has been estimated at $700 billion by 1980, including new stocks and bonds of about $250 billion annually. The total world market value of outstanding stocks and bonds could then exceed $3 trillion.[145] The sums the Arab nations may have left over, after internal and regional development, will not enable them to buy much of America. As we noted earlier, the more likely and more difficult problem is that the Middle Eastern nations will limit oil production because such investments hold small attraction to them.

The U.S. is certainly being affected by higher oil prices but if we hold oil imports to the current 15 percent of our energy supply it is a manageable problem. Indeed, the impact on the U.S. is much less severe than on many other consuming nations. The higher prices makes it much more likely that domestic alternatives can be economically developed to keep U.S. imports from getting out of control.

It is not the U.S. but Europe, Japan and, even more, the Third World, that is bearing the economic brunt of the new shortages and high prices of energy. Throughout the past 20 years Western Europe and Japan have steadily enlarged the share of their energy consumption supplied by oil from the Middle East. The oil has been sold to their consumers largely by American owned international petroleum companies with production in the Middle East and distribution facilities in the consuming nations. As the North Sea oil is developed in the years ahead, Great Britain and Norway will become largely self-sufficient, but for Western Europe as a whole and for Japan the next decade will mean a continuing heavy dependence on oil from a handful of Arab nations along with Iran, Nigeria and Indonesia.

The apprehension of the Europeans and Japanese is heightened by the fact that they are dependent largely on American-owned companies such as Exxon, Mobil, Texaco, and Gulf to sell them the Middle Eastern oil they need. The Europeans and Japanese may rightfully wonder whether in a real crunch, such as arose in 1973-74, they obtained their fair share of the available oil supplies as compared with the U.S. The fact that production in the nations with the largest oil reserves, such as Saudi Arabia, is entirely in the hands of American companies adds to the apprehension of other consuming nations.

Even before the crisis of 1973-74, growing U.S. imports of petroleum had helped push up the prices paid by the Europeans and Japanese. The most recent doubling of prices means that the European and Japanese economies will feel a real pinch. The consequences are much more troublesome than mere price increases. The Japanese especially are undergoing a basic retrenchment of their once-booming economy. The energy *dolce vita* has come to a sharp and painful end. Unless Europe and Japan learn how to trim their energy demands and develop alternative sources they will find themselves bidding desperately against one another and the U.S. for their share of an inadequate supply in the 1980's and beyond.

The quantum jump in the producing nations' income from their sales of oil—from $1.50 per barrel in 1973 to $7 in 1974—will result in a significant redistribution of wealth among the nations of the world. Japan and Western Europe will become somewhat less rich, the oil producing nations will become quite rich and the billions of people in the Third World who produce little or no oil will become even poorer. The problem of readjustment for Japan and Western Europe will substantially drain off their financial capacity to assist the Third World and weaken their demand for imported raw materials.

The picture is truly dire for the energy consuming countries of the Third World. As prices soar and the large consuming nations scramble for available supplies, the smaller nations without indigenous energy resources are finding it increasingly difficult to develop modern economies and improve the lives of their people. Food production is curtailed in India, for example, because it simply does not have the money to pay the higher price of fertilizer, which must reflect the high cost of petroleum from which it is made. Of course, for nations with surplus petroleum or uranium the growing world demand for energy is a key to prosperity. But for the billions of people who are poor and live in lands without energy resources the escalation of energy prices is extremely bad news.

It is the Third World that will be the prime victim of the massive flow of wealth to the Middle East unless the OPEC nations share that wealth to some considerable extent. The OPEC nations are now charged with the responsibility of power. They can become true world leaders by taking steps to assure that the poorer nations can obtain the oil they need at prices they can afford and can pay through loans or outright grants. By foregoing a small percentage of their new wealth which they do not need the oil producing nations can set an example for the still very much richer Western world.

The result might well be a more vigorous common effort to end starvation and dire poverty in the world.

It is no exaggeration to suggest that solution of the energy supply problem is essential to world peace in the decades ahead. In this country we are already rationing petroleum to assure that farmers, industry and other priority consumers don't run short. But there is no international rationing board to protect small nations, nor is there likely to be one. If energy supplies remain expensive and scarce, the industrialized consuming nations are likely to obtain their share but the poor nations will be left out in the cold. By the turn of the century, seven billion people throughout the world will be aspiring to participate in a high-energy civilization. They will either become significant consumers or foment turmoil and revolution that will disturb world peace.

These problems raise the issue of what form of international collaboration we should be seeking. The U.S. has taken the lead in urging closer cooperation among the consuming countries as a response to the remarkable solidarity shown thus far by the producing countries. But it is clear that the consuming nations can accomplish very little by forming a block to exert countervailing power. Differences between the U.S. and other nations in relative dependence on imports and foreign policy in the Middle East make it difficult to agree on a program. More fundamentally, the producing nations are indispensible to any solution.

Collaboration must go beyond the issues of price, the major concern at the moment, to include joint research and development and investment to open up alternative supplies and programs to conserve energy and reduce future consumption. Collaboration should also include the programs of technical and managerial support for the Middle East previously discussed and special solutions for developing nations. The worldwide energy problem provides a fresh opportunity for collaboration through bilateral and multi-lateral arrangements. The goal must be to shape a solution acceptable not only to the large consuming and producing nations but "future consumers" as well.

What happens in America, the world's largest importer of

crude oil and its products, is naturally of vital interest to the international market and in particular to other consuming countries. Thus one major element of a solution is to scale down the rate of growth in overall energy consumption in the U.S. and shift the growth away from imported oil as much as possible. The link between conservation at home and adequate supplies for the world at large is a direct and vitally important one. Energy conservation could release supplies for the use of other consuming countries and would buy time for developing alternative sources. It could avoid the worst aspects of a foreign policy and monetary dilemma that otherwise faces all the consuming nations, large and small, if we continue to slide into greater and greater dependence on Middle Eastern oil. By now we know that action by any nation appreciably to reduce wasteful energy consumption or to increase supply relieves the problems of all other nations. For there is indeed a world market in energy. We are all in the same boat. But most of the people in the boat are still energy-starved.

Energy conservation can reduce the U.S. drain on the world energy supply. But from a worldwide perspective, the only way to avoid a worsening energy crisis is to enlarge the supply of cleaner energy. At the moment each of the major consuming nations is pursuing independent research and development programs toward that end. But amazingly they are all devoted largely to atomic energy and essentially duplicate one another. The United States, Great Britain, France, West Germany, the Soviet Union and Japan all have large-scale programs to develop the atomic breeder reactor and they are all pursuing the same technical option, the liquid metal fast breeder. Most of these nations are also heavily engaged in research into fusion power. Of the major industrial nations only the Soviet Union and the United States have substantial research programs outside the field of nuclear energy; the United States has only begun to mount a diversified effort.

The energy source that most of the peoples of the world have in common is the sun, but the neglect of solar energy research and development is also a common feature of the international picture. The other options for cleaner energy also have been largely ignored around the world. The new awareness of environmental pollution has caught the energy establishment in other nations as flat-footed as in the United States.

Perhaps industrial rivalry explains the multi-billion dollar worldwide duplication of effort in nuclear fission research. Yet a worldwide high-energy civilization should not have all its chips riding on one number on the roulette wheel, even if it does offer better odds than other choices. On a worldwide basis there is an urgent need to pursue alternative breeder concepts, solar energy, geothermal and other potential sources, as we will discuss in chapters 11 and 12.

It would amount to economic madness for the nations of the world to follow the current pattern of duplicate programs in developing solar energy and other new sources. The fact that the research slate is still quite clean is an advantage in the sense that no nation has a lead to surrender. The financial and technical talent of both the industrial and developing nations should be pooled in an international effort to enlarge the supply of cleaner energy. It is important that the developing nations be directly involved in the research effort so that the program includes technical options in small enough packages to suit their needs. The nuclear fast breeder reactor may turn out to be fine for the United States and the Soviet Union, but nations like Cameroon, Bangladesh, Colombia and Burma need their energy supply in smaller packages than the million kilowatt power plant.

A research and development program for solar energy offers a unique opportunity for international cooperation to provide the array of clean energy technologies needed throughout the world. It is an especially promising option because the vast majority of the people on earth live in the temperate zone where the possibilities for harnessing solar energy are greatest. The oil producing nations in the Middle East have indicated an interest in investing large sums in such an effort. Their deserts are prime sites for solar energy stations that would replace their depleting oil and assure the future of the energy intensive economies they hope to build.

And as we shall see the concept of a satellite solar station in space is uniquely adapted to becoming a simultaneous source of power for many nations.

Joint funding of these research programs could create the critical mass of talent that would shorten the lead-time until they become available. One need not be naively optimistic about international cooperation or overlook the virtues of competition and diversity in the research field. An international program should encourage great diversity in the research effort while ideas are still in the "test tube" stage. There would be the stimulus of communication and competition. But what could be avoided is duplication in the more expensive building of large scale demonstration projects. To achieve such cooperation will require an acceptance of the urgency of the problem and the interest and determination at the highest levels of national governments. Success would at once help solve the problems of energy policy while enhancing international relations in many other areas of common concern.

The energy crisis is also the environmental crisis. The 1972 Stockholm Conference on the Human Environment demonstrated the international nature of the pollution problem, but what has not been sufficiently accepted as the central role of energy production in causing environmental pollution. The implementation of the Clean Air Act in the United States depends on the availability of clean energy and the same connection between standards and supply exists on a worldwide basis. If the nations of the world will join hands in cleaning up the environment, they are bound to join hands in developing cleaner supplies of energy.

Some of the environmental threats raise serious problems of foreign policy. The need to safeguard nuclear fuel against diversion into nuclear weapons by individuals or governments is a most worrisome problem. No nation can really be safe from the threat of an atomic bomb made from the waste products of a nuclear power plant unless there are safeguards enforceable in all nations. The nuclear threat is only the most dramatic of the international pollution problems associated with energy production. All of the damage to the U.S. environment arising from energy production, of course, also occurs in other nations. And the pollution of the oceans and the air ignores national boundaries.

Simply exporting the pollution problem cannot be a long-term U.S. policy. If offshore drilling is too uncertain to risk U.S. beaches and the stripping of land is too ruinous to justify mining coal or shale oil, then it is not realistic to expect to obtain these energy sources from abroad for any extended period of time. Placing power plants in the desert in the Four Corners region of the southwestern U.S. did not solve the energy-related pollution problem. And neither will the problem be solved by locating polluting industries in the Bahamas, North Africa or even the Sahara Desert.

Developing nations may initially choose to accept pollution as a price for economic progress using current technology just as this country did in an earlier era. But the U.S. cannot expect to continue a high-energy economy on the basis of fouling other people's air and water and destroying their land. Even if the planet could survive, improvements in standards of living mean that the time would soon come when people in these nations would demand the same respect for their basic resources that sustain life as now dominates American thinking.

Thus as we envision efforts to assure the adequacy of energy supply on this planet, we realize that it must also be a clean energy supply—one which will not impair the physical or mental health of the entire earth's population.

The international aspect of energy policy is much more than concern over worldwide shortages, foreign policy constraints and monetary instability. Any place on earth where energy is produced, shipped, or consumed we see many serious threats to the environment. And where energy is not extensively used we see the poverty from which most of mankind has yet to escape. An adequate supply of clean energy for the entire world is thus an absolutely necessary prerequisite to satisfying the desires of all of the people on earth for a better, more peaceful life. Future energy policy is therefore synonymous with a sensible policy for the future of civilization.

CHAPTER 7

The Price of Energy

To consumers, the most important and alarming fact about energy prices is that after two decades at bargain levels they have now skyrocketed. This is a jolting reversal for a nation that has taken low energy prices for granted. And the prospect is for continued price increases in the future. Popular concern over these trends is heightened by a growing concentration and growing profits in the basic fuel industry, which suggests that competition may not be adequate to assure reasonable prices to the consumer.

The oil producing nations of the world "dropped the other shoe" early in 1974 with an oil price earthquake that shook the entire consuming world. The posted price of crude oil—f.o.b. the Persian Gulf—which had already climbed from $1.80 in January 1970 to $5.12 by October 1973 was doubled overnight (to $11.65). Crude oil that was delivered to the East Coast of the United States for $2.25 in 1970 now costs at least $9.00 a barrel.

These enormous price increases have converted the world oil market from a competitive force that kept a lid on U.S. fuel prices to a giant balloon pulling every other price up up and away with it. Domestic oil prices, formerly protected from the world market by the quota system when world prices were lower, are now being permitted to rise and approach the escalating world prices. Oil itself is almost half the energy story, but with the oil industry's entry into coal and uranium all fuel prices are reaching for new highs.

Consumers now spend about $80 billion a year in direct purchases of energy, on utility bills, heating fuels and more than half on gasoline and oil for the family car. In addition energy purchased by industry forms a part of the cost of almost everything else we buy. Sometimes this part is very small, as in most items of furniture; in other cases, such as aluminum foil, it is substantial. On the average, however, energy represents perhaps 4 percent of the final cost of most products we buy and use.[146]

The cost of energy adds up to an important item in the family budget, especially for those at the lower end of the income scale. For the average family, direct purchases of energy for use in the home and for the car amount to about 8 cents out of every dollar spent; if we include the energy component of everything else we buy, the cost rises to about 10 cents out of every consumer dollar—as much as the average family spends on clothing.

For low-income families, energy takes an even larger share of the budget. Even though they own few energy consuming devices, they may spend as much as 15 percent of their money on energy. For families in the high income brackets 5 percent is probably more typical.

Until very recently, the price of energy has been one of the great success stories of America. For example, over the past 40 years the price of electricity has halved, while prices in general have doubled. As President Nixon stated in his Energy Message of June 4, 1971, "Energy has been an attractive bargain in this country—and demand has responded accordingly." Energy prices remained stable in the 1960's, resisting the inflationary trend of those years. Consumers increased their consumption of energy without feeling any pinch because the prices of everything else, and their incomes, were rising faster than the price of energy. From

1954 to 1972, expenditures on energy thus remained between 7.1 percent and 7.6 percent of the average consumer's total budget despite a 50 percent increase in per-capita energy usage.

There are sharp regional variations in energy prices. Petroleum producing states such as Texas enjoy relatively low fuel prices. Energy prices are highest in New England and to a lesser extent the Middle Atlantic states which pay more than 25 percent above the average national price for electric power and nearly double the average national price for natural gas. The bargain states for electricity are those in the Tennessee Valley and the Pacific Northwest where rates average better than 33 percent below the national average.

The nation's use of electric power and other forms of energy is very uneven. Air conditioners operate only a few months in the summer, and heat is used only in winter. Most businesses are shut down at night. Yet the utilities must build enough power plants to meet the peak use—the hottest or coldest day of the year. But almost half the time the nation's power plants stand idle.

It is important to recognize that the pricing of electricity, and other forms of energy as well, can be designed to encourage savings. Discounts for electricity used at night would induce some industries to operate after dark. Charging a higher price for consumption during the peak hours would encourage consumers to invest in more efficient air conditioners and in improved insulation of their homes. Such pricing techniques would cut down demand at peak periods and year-round consumption as well.

The following brief survey of the main sources of energy indicates recent trends in retail prices and their impact on consumption and expenditures. Recounting the favorable experience of the past is useful to help us appreciate more fully the impact of the problem which most Americans are beginning to feel, and one which they will feel more intensely in the future.

Electricity—the most important form of household energy in terms of expenditure—is sold on a sliding scale that reduces the unit price as the volume a customer uses increases.

A typical rate schedule for a residential customer, for example, charges 4.5 cents per kilowatt hour for the first 100 kwh per month, 2.75 cents per kwh for the next 150 kwh per month, and 2.0 cents per kwh for the next 250 kwh. As families increase their monthly purchases of electricity they pay a lower price for each additional kwh.

As a result of the sliding scales of rates, the average price for electricity paid by residential customers served by private companies fell from 2.9 cents per kwh in 1950 to 2.2 cents per kwh in 1970.[147] The sliding scale rates have also been very successful in promoting increased use of electricity. Average per-capita consumption increased almost 400 percent from 1950 to 1970.[148]

The decline in the average price of a kwh of electricity from 1950 to 1970 continued a much longer term trend. From 1926 to 1968 the average unit price for electricity paid by consumers fell by almost half while consumer prices generally rose by about 100 percent.[149]

The price trend for electric power has now reversed itself. Since 1968 the average price of electricity has increased about 5 percent each year until 1973-74 when it jumped up about 20 percent. An estimated $1.7 billion-worth of rate-increase applications—in all-time high—were pending before the various state regulatory agencies.

The retail price of a cubic foot of *natural gas* rose fairly sharply from 1950 to 1960 and then, like electricity, remained stable until 1968. Average consumption of natural gas by households doubled between 1950 and 1968[150] as more and more homes converted to this clean, low-priced fuel. During this period natural gas bills rose by only 25 percent despite a 44 percent increase in consumer prices.

Again as with electricity, natural gas prices vary considerably between different regions of the country. In a typical month in 1973, for example, the average price per 100 therms (the residential heating rate unit) varied from $6.39 in Houston, near the gas fields, to $17.84 in Boston, where prices include the cost of transportation from the Southwest.

The 1960-68 period of price stability for natural gas has now reversed itself with an increase in retail prices of more than 5 percent per year since 1968. Natural gas prices will continue to rise sharply. New supplies at the wellhead or from abroad are often double or triple the price of older supplies already under contract.

Prices of *fuel oil* for home heating purposes have followed a somewhat different pattern. Fuel oil prices show considerable year to year variation, but in contrast to electricity and natural gas, fuel oil's price went up in line with prices generally in the 1960's. Fuel oil consumption during these years increased only slowly. Unlike natural gas and electricity, fuel oil prices appear to show little variation between different parts of the country. From 1968 to 1972, the price of home heating oil has also increased rather sharply at about 5 percent annually in the eastern United States and somewhat less in the nation as a whole. At the wholesale level, heavy fuel oil prices increased almost 70 percent from 1968 to 1972, primarily as a result of a shift in demand from high sulfur oil to low sulfur oil. But in the winter of 1973-74, fuel prices really took off, soaring another 50 percent.

As noted previously, the average consumer spends more on energy for his car than he does on all forms of household energy. The total price of *gasoline* rose generally in line with the cost of living up to 1968. But the primary increase was the result of rising state and federal gasoline taxes. Gasoline prices were relatively stable from 1950 to 1968, increasing only 14 percent. On the average state and federal taxes on gasoline (used mainly to build new roads) increased from 7 cents to 11 cents during the 1950-68 period.

Compared with all other forms of energy, the price of gasoline remained relatively stable during the 1968-72 period with an overall increase of less than 2 percent per year, much less than the consumer price index. Even this modest rate of increase was much faster than in the previous ten year period. But as we all now know gasoline has now been caught in the upward spiral with a vengeance. The average retail price of gasoline jumped from about 30 cents per gallon at the beginning of 1973 to 50 cents per gallon by March of 1974 and seems destined to go higher in the absence of controls.

Inter-fuel competition seems to have played a beneficial role for consumers during the 1960's. The retail prices of natural gas and electricity remained stable, and these fuels increased their share of residential energy consumption. The increase was particularly notable for electricity. The price of fuel oil rose more sharply and its share of the household market shrank. At least in the past when sources of supply were abundant, inter-fuel competition has proved to be a powerful consumer protection force.

There is, of course, a limit on the influence of relative prices on various sources of energy. They do not compete solely on a cost basis; if they did, electricity, which is by far the most expensive form of energy ($7.50 per million btu compared with $2.25 for fuel oil and $1.25 for natural gas), presumably would be used much less than it is now. Electricity has a virtual monopoly on uses such as household lighting. For heating and other uses, factors such as cleanliness, convenience, the cost of heating systems, and the shortage of alternative sources are factors which also affect the choice. Comparative prices may not always control, but over time they can play an important role in deciding which form of energy consumers prefer.

The reasons for stable energy prices in the past are clear. In the post World War II period the U.S. still had at its disposal huge quantities of readily accessible fuels that were very economical to produce. Half the growth during this period was supplied by natural gas from the Southwest, found largely as a by-product of the search for oil. Large blocks of coal in thick seams were still readily available for power plants located near the big markets in the East, and there were no environmental constraints on their use. Crude oil was so plentiful in world markets that from 1959-73 the federal government imposed import quotas to protect the domestic oil industry from competition.

Backed by this rich array of ready reserves of fuel, the energy industries pushed their technological advances to their ultimate. They employed the mass production techniques and promotional pricing concepts first demonstrated by Henry Ford. And the energy companies as well as their customers benefited. The fuels were brought to market through the rapid development of innovative technology in all branches of the industry—oil exploration, drilling, pipelining, and refining; electric generation, transmission, and distribution; and coal mining and transportation. All energy industries took full advantage of cost-reducing economies of scale.

Productivity increased enormously. From 1950-68, output per man hour in all the energy industries, including coal, rose at a much faster rate than in the private sector as a whole, and in all cases outstripped the rise in average hourly earnings. To cite some examples of technological improvement: An electric power plant in 1968 could generate a kilowatt hour of electricity with 0.87 pounds of coal while it took 1.19 pounds in 1950.[151] Petroleum drilling techniques were improved after World War II so that the cost of drilling per foot remained stable in the face of a 23 percent increase in material costs during 1965-70 alone.

A quick analysis of the cost structure of the energy industry gives an insight into these price trends. An important fact often overlooked is that the basic fuel represents only a small fraction of the ultimate price of energy to the residential consumer. For example, only about 20 cents of the dollar spent on electric power by the ultimate consumer is needed to pay for the power plant fuel. The generating plant itself costs another 30 cents and about 10 cents is required to cover the cost of the large transmission lines from the power plant to the local distribution systems. About 40 cents of the consumer's dollar is required for the distribution of the electricity within a metropolitan center including the home meter reading and customer service.

Similar patterns hold true for other forms of energy. Almost 70 percent of the delivered cost (excluding taxes) of natural gas to a residential customer, for example, is accounted for by costs of distribution within the local area. In 1973 the wellhead price of natural gas was only 15 percent of the price the consumer paid.[152] For heating oil, the cost of refining and distribution to the consumer represented 60 percent of the retail price. In 1973, production costs represented only 30 percent of gasoline prices excluding taxes, and an even smaller part—20 percent—if taxes were included in prices. Even today with the escalation of crude oil prices only about 15 cents of the 50 cents the motorist pays for his gallon of regular grade gasoline is needed to cover the cost of producing the crude oil. The great bulk of the cost, as in all forms of energy is in conversion, transportation, and distribution within the marketing area.

The importance of distribution costs, particularly the final distribution stage to consumers, helps explain why large industries can buy energy at much lower prices than a household consumer. There are also significant economies in selling larger quantities of electricity and natural gas to each householder since the pipes, power lines and customer service are required in any case. The economies explain the cost basis for rates which decline as the volume purchased increases.

But no system of pricing can conceal the effect of higher prices on consumers, small and large. Why is it happening? Is the energy crisis being used as an excuse to justify rising prices and higher profits for the companies?

Unfortunately, higher energy prices arise from fundamental causes, such as higher costs of fuel, labor and capital and a slowdown in offsetting efficiencies. Stopping the energy price spiral is much more difficult than simply trimming corporate profit margins even though windfall profits by the large energy-producing companies are running into the billions each year.

The energy industry has done quite well financially over the years. No major oil company, electric power company or natural gas pipeline or distribution company has gone broke in the past 20 years. Rates of return in the oil industry fell slightly in the 1950-68 period and actually increased somewhat for the gas and electric utilities. Profits in the coal industry are less easy to measure but in marked contrast to the other energy industries, these fell sharply in the 1950-68 period[153] as coal lost out in the inter-fuel competition.

The rates of return on investment of the regulated gas and electric utilities appear to have actually declined in recent years as costs increased and rate increases lagged behind because they required regulatory approval. But the energy-producing companies are not passing up the opportunity to fatten their profit margins.

In 1973 oil company profits began to rise rapidly and were 50 percent higher than 1972. Return on the stockholders equity in the 15 major companies jumped from under 11 percent to over 14 percent. And 1974 looks like another year of extraordinary profits. The coal industry, particularly the larger firms, has also increased profit margins in recent years.[154] Ownership of many of the large coal companies has passed into the strong hands of major petroleum companies. These mergers with oil companies plus a sharp increase in coal prices in the past few years (an average of 18 percent annually since 1968) may help explain why their profit margins now resemble those of oil companies rather than the smaller profits that the highly fragmented and competitive coal industry earned in the past.

Nevertheless, rising energy prices far exceed the margins that might be traced to higher profits. The explanation is that the nation has depleted its lowest-coal fuel resources, and the steady flow of technological innovation has run dry. The remaining fuel sources, particularly domestic oil and natural gas, are more costly to locate and produce. We have to go deeper and further for additional reserves. During the last decade both the industry and government have neglected research and development that might increase the efficiency of existing energy supply systems or provide new sources of energy. And of course a major source of the most recent increased costs is the higher royalties that the producing nations are receiving for their oil.

Thus, new increments of fuel supply to meet growth in demand are more costly than ongoing supplies. And with few increases in productivity in sight, new power plants and refineries to convert the fuel must bear the brunt of the high inflation in the cost of materials and wages.

The energy industries are uniquely capital-intensive. It takes a $5 investment for every $1 of electric power revenue, and capital requirements for the other energy sources are almost as large. This means that construction costs are a dominant factor, and these rose by more than 30 percent between 1965 and 1970. To the inflation in actual costs of construction must be added the higher interest rates at which capital can be borrowed, which are also a very important item of cost, especially to the electric power and natural gas companies. The interest rates in the 1970's are in the 8 percent range, 50 percent higher than those prevailing in the past two decades. As a consequence of all these factors, an electric power plant under construction today will produce electricity at the plant for roughly 9 mills per kilowatt, which is almost twice the unit cost when a new plant began operations a decade ago. New supplies of natural gas are also at least twice as costly as gas now flowing to consumers.

The cost of distribution of energy within the metropolitan centers—which amount to almost half the total retail price—is following a somewhat different pattern. Costs should continue in the same trend as in the 1960's. True, the cost of labor and materials will increase but the systems are by and large already in place for electricity and natural gas and there are still some economies of scale as existing customers continue to increase per-capita consumption.

The increased price of energy to the consumer thus stems largely from the increased cost of fuel and its conversion to electricity or gasoline and fuel oil, and not from increased profits. But this is small comfort to the consumer. The public could well inquire why energy industry management seems unable to bring runaway costs under control. Some of the costs are beyond their control but they are influenced by the fact that extra costs can be passed along to their captive consumers. We must question the regulatory system that fixes rates on a cost-plus basis, a system which provides no incentive for management to cut costs or to undertake research and development for more efficient, cost reducing technologies.

The new laws to protect the environment and the health

and safety of workers are also adding to the price of energy. It is important to realize that these measures are not adding new costs to society, unless they are ineffective. They shift to the price of energy and therefore the consumer the burden of controlling wastes that are otherwise much more costly in damage to human health and the natural environment.

Nevertheless cleaning up the environment will be an increasingly significant factor in causing the price of energy to rise. New health and safety measures are resulting in some increased costs of mining in order to protect the lives of coal miners. New regulations to minimize oil spills in offshore drilling are adding to the cost of oil and associated natural gas production. Air pollution controls require the use of low-sulphur fuels that are 50 to 100 percent more costly than the dirtier fuels that were formerly used. And if the dirty fuel is still used, pollution control equipment required for the power plants and oil refineries add appreciably to their cost. Power lines in many new suburbs and scenic areas are placed underground, which is more expensive than using overhead lines. And the need for better planning and environmental reviews has stretched out the time required to complete new power plants and other construction. With an eight year lead time, the interest during construction on a new power plant adds over 20 percent to the cost.

These environmental costs are difficult to quantify with precision, but, on a per-unit basis, they account thus far for only a small fraction of the increased retail price of energy. The percentages vary. For natural gas, environmental protection costs are very small. For oil they are also small and over the years will not exceed some 10 percent of the retail price of gasoline. For electricity, environmental costs may account for 10 to 15 percent of the retail price even if coal is the fuel. Yet in aggregate these sums are large: An added ten percent on the U.S. energy bill, which by 1980 may be $200 billion, would be $20 billion per year. It is therefore important to protect the environment as efficiently and economically as possible.

Whether we can avoid soaring energy prices in years ahead depends on providing the industry with sufficient incentive for efficiency. It will require a greater stress on quality control so that consumers do not continue paying for plants that are broken-down much of the time and on lower-cost construction methods to stop the trend of cost-overruns in the construction of energy supply systems. It will also be necessary to assure that profit margins remain reasonable, which means in line with earning levels in the 1960's.

The prospect of a continued rise in costs and prices focuses more attention than usual on the market forces and government policies that might encourage efficiency and keep both costs and profits in line, namely, the competitive conditions and framework of regulation under which the various energy industries function.

We can be sure that where monopoly power exists, the producer will raise prices to whatever the market will bear to maximize profits. In such a case price controls are necessary if private enterprise is to serve the public interest. At the other extreme is the condition of aggressive competition where there is a strong incentive to limit profits to the minimum needed to stay in the market. Such competition also exerts a great pressure to reduce costs.

Electric and gas utilities serving a franchised area have always been recognized as natural monopolies requiring price regulation. And the interstate pipelines are subject to federal regulation because most of their customers are captive to a single pipeline company or are served at most by two or three pipelines that operate as natural monopolies. But the production of oil, natural gas, coal, and uranium is quite different.

Fuel production has few of the characteristics of a natural monopoly. Yet one can hardly claim petroleum producers are competing for sales of crude oil or natural gas on the basis of which company offers the lower price. The competition seems rather to be among the buyers scrambling for whatever supplies can be purchased. Fuel producers occupy an uneasy position in the middle ground between competition and monopoly.

The economist's definition of perfect competition is rarely fulfilled in daily economic activity. It is therefore customary to inquire whether conditions in a given industry are "workably," rather than perfectly, competitive. This depends on several key aspects of the industry's structure including the domination of the market by one, or a few, sellers, the ease with which potential competitors can enter the industry, and the possibility of consumers "buying something else" if the price of the product goes too high.

The issue is well illustrated in the current debate over deregulation of wellhead prices in the natural gas industry. The first question to be answered is whether market forces can protect the consumer in the absence of price controls. A brief look at the structure of the present day natural gas industry is necessary for the discussion. That industry is made up of three very different kinds of private companies. Production is primarily by the major oil companies and a large number of independent producers. About 100 pipeline companies—most independently owned—buy the natural gas from the producers, transport it by pipeline and sell it to the local distribution companies or directly to large industrial consumers. Some pipeline companies also produce natural gas. The local distributors of natural gas are the descendents of the old manufactured gas companies. They compete with the electric utility companies (when they are not the same company) as well as the local fuel oil dealers.

All three branches of the natural gas industry are made up of privately owned corporations except for municipal distributors of natural gas, which account for less than 10 percent of the industry. There is fairly comprehensive public regulation, however. The Federal Power Commission regulates the prices at which interstate (though not *intra*state) pipeline companies may buy gas in the field and sell it to their retail distributors.

There is a considerable degree of concentration in the production of natural gas. Sales by natural gas producers to interstate pipelines (which probably cover between 60-70 percent of total sales) indicate that the top twenty petroleum producers have been accounting for between one-half to two-thirds of total sales and their dominance has been steadily increasing. More important, however, is the fact that production is regionalized and a few producers typically dominate a particular region.

The pipelines are tied to the producing territory in which they are located and are faced with demands from their customers for larger quantities of natural gas than they can purchase. In a real sense they are at the mercy of natural gas producers because, without gas to transport, they face the prospect of a sunk investment lying half idle in the ground. The pipeline company's primary incentive is to maximize the flow of natural gas through its system. The cost-plus regulatory formula means that they can pass on to their customers the higher prices charged by producers.

Notwithstanding the fairly high degree of concentration among producers and the poor bargaining position of the pipelines, most economists agree that "workable competition" does exist among gas producers.[155] Their conclusion is based primarily on the number of companies in the business including the large number of small producers—gas and oil producers are still numbered in the thousands—alongside the giants. Furthermore, the barriers to entry in the natural gas producing business are said to be substantially less than in most manufacturing, and if the high risks attached are most easily absorbed by large companies, they are also attractive to small concerns, including individuals in high income brackets.

Unfortunately the number of small producers is declining rapidly.[156] More important they are confined to onshore drilling, where the best oil and gas prospects have long since been developed. Under prevailing leasing policies only the major producers have the capital to participate in the offshore areas. Thus entry is really not very easy at all, insofar as offshore natural gas production is concerned, especially if the government continues to award bids on a cash bonus basis which effectively excludes smaller companies.

Actually the fact that there is "workable competition" among gas producers by an economist's definition misses the reality that concerns consumers. The crux of the problem is the superior bargaining power of the producers, no matter how numerous they are. Demand for natural gas is much greater than supply, and all producers have the power to boost prices. And higher prices for natural gas over the past two or three years demonstrate that the shortage is caused by factors other than price. Environmental constraints in the pace of drilling and all the other constraints discussed in Chapter 5 are at the heart of the natural gas supply problem.

The policy issue concerning price controls on natural gas cannot be settled to the satisfaction of the public by the textbook test of whether or not the structure of the industry is competitive.

The petroleum shortage is likely to continue as long as major producers such as Saudi Arabia and Libya limit production. And no one can claim that prices reflect price competition as long as these nations with monopoly power fix the world price of petroleum and the American major companies follow the crest of the wave. Given a free market the price of natural gas, which is still largely a by-product of the search for oil by the major companies, would escalate to a "fair market value." In practice, this means the level of the alternative fuel—oil—which in the U.S. is soaring to the levels set by the Arab oil producing monopolists.

There may be a basic distinction to economists between monopoly profits and a situation where the industry is competitive in structure but is selling a product so much in demand that it can earn monopoly-sized profits. To the consumers who pay the soaring prices, the difference is academic. The policy issue that will be discussed in Chapter 13 actually turns on whether consumers have no choice other than to pay these windfall profits or run short of energy.

The same companies that produce natural gas are also the oil producers. Insofar as competition is concerned oil from other nations is now determining the price for domestic production. For decades, U.S. crude oil prices were supported by oil import quotas and production controls that shut out foreign competition. Now oil can be freely imported. Yet the embargo, a tight supply throughout the world and foreign policy considerations raise doubts about the vitality of oil competition. The producing nations are now capturing the enormous difference between the per-barrel cost of producing Middle Eastern oil—measured in pennies—and the price at the wellhead, which is all the traffic will bear. The international petroleum market has been dominated until recently by a handful of large international companies, mainly American. These companies, through their control of Mid-East production, for decades until the recent price revolution by the producing nations, were able to capture for themselves as profits much of the difference between the low cost of production in the Middle East and the high selling price in the U.S.

Recently the producing nations have turned the tables on the international oil companies. The American consumer is between a rock and a hard place. The tussle between the international oil companies and OPEC has been over the division of the economic spoils: OPEC now seems to be the winner but the oil industry is by no means the loser; their profits are higher than ever. It is the American consumer, and consumers in every nation, that are the real losers. And there seems little prospect that competitive pricing will soon return to the world scene despite the hopes and predictions that the OPEC cartel will disentegrate.

Competitive conditions in the domestic industry are also in need of strengthening. The large integrated petroleum companies are well on their way to becoming vast "energy" companies, encompassing all forms of basic fuels of both current and potential importance. This development has important implications for lessening inter-fuel competition and therefore overall competition within the energy industry. During the 1960's, several large integrated petroleum companies merged with each other and acquired assets of independent marketers, refiners, and producers.[157] The coal industry too, which for much of its long history was said to be reasonably competitive, experienced a dramatic change in its structure in the 1960's. A series of mergers resulted in greatly increased concentration in the industry. The thirteen largest companies accounted for 51 percent of total coal production in 1969 compared with 39 percent only seven years earlier.[158]

The effective degree of concentration is even greater than

indicated by these figures. Part of coal production is captive, i.e., owned by consuming companies such as steel, and does not enter the commercial market. Furthermore, some of the largest companies market the coal production of other companies in addition to their own. Taking these two factors into account, it is probable that the thirteen largest companies control about two-thirds of the coal sales in the commercial market. The growing concentration of ownership in the coal industry in itself would not be alarming. But the ownership is shifting to the same companies that already control oil and natural gas. And that does pose a real threat to interfuel competition.

The major oil companies played a large role in the coal industry mergers. By 1969 the oil industry accounted for almost a quarter of total coal production compared with a mere 2 percent in 1962. The steel, copper and other industries also control substantial coal production. In fact the independent coal companies' share of coal production fell from 32 percent to 10 percent in the 1962-69 period.

To be sure, it is not necessarily a restraint of trade for an oil company to penetrate the coal market. If it creates a new company in the coal business, the number of competitors is increased and could be a stimulus to more vigorous innovation and efficiency, translated into lower prices. But when the oil companies enter the coal business by taking over existing large coal companies, as has occurred, or when the integrated energy company represents a significant share of the market in both fuels, there is a real danger that inter-fuel competition will disappear.

Inter-fuel competition was an important market regulator in the 1960's. Even if the coal, oil, or uranium industries were not highly competitive within their separate industries, the growing competition among them resulted in fair prices to consumers. Inter-fuel competition flourished while ownership of the different branches of the energy industry remained distinct. But this is no longer so. The petroleum companies have not only moved into coal; they have acquired coal lands[159] and are expanding rapidly into all branches of the energy field. Of the 14 largest petroleum companies (ranked by 1969 assets) seven (including the four biggest) had diversified into all other branches of the energy industry—gas, oil shale, coal, uranium and tar sands.[160] The other seven companies produced oil, natural gas and at least one of the other forms of energy. It has been argued that the extension of petroleum companies into synthetic crude oil from coal is technically logical because the oil thus obtained would be processed in a refinery in the same way as natural crude oil. This may be true but the coal companies acquired by the petroleum industry are selling coal as usual. Whatever the motives may have been for the oil companies' diversification into uranium and coal, the result appears to lessen inter-fuel competition.

It could well be that these energy industry mergers and the growing concentration of ownership do not violate traditional anti-trust principles. But the threat to the public interest is nevertheless real. It is no accident that the National Coal Association and the petroleum industry now sing the same tune. The political and policy influence of this greater concentration of corporate wealth can easily be overlooked.

The problems posed go deeper than simply prices. The waning competition within the energy industry poses a threat to the adequacy of supply. One does not have to subscribe to a conspiracy theory to observe that a shortage of energy is a situation most favorable to the energy companies. Concentration within the industry makes it less likely that any producer will scramble to enlarge supplies unless the profit margins are most attractive. The energy companies dig, drill, or build only if they believe the price is right. Government price controls will be more necessary than ever if inter-fuel competition is permitted to wane.

The shortage of energy is already sapping the strength of competition in the retailing of gasoline. Independent stations rely on the major companies for their supply because the majors control some 90 percent of gasoline production. Naturally the big companies have cut short their competitors while continuing to supply their own stations. As a result, independent stations have gone out of business by the thousands since the summer of 1973. And those that have gasoline for sale are certainly not cutting prices.

The hard fact for consumers is that there is no price relief in sight. There is every reason to believe that the prices of all forms of energy will increase more rapidly than the cost of living throughout the decade of the 1970's. Even if inflation were to be brought under reasonable control, the retail price of electric power and natural gas in 1990 is projected to be at least double current prices. The Federal Power Commission in its *1970 National Power Survey* estimates that costs of electricity to the ultimate consumer could rise from 1.54 cents per kwh in 1968 to 3.51 cents per kwh in 1990 if the same rate of inflation which was experienced in the 1960's (3.0 percent a year) were to continue in the future. Moreover, this estimate includes the savings in the sliding scale as volume of use increases.

For anyone living in even a modest-sized all-electric home, the prospect of finding a $100 electric power bill in your mailbox for a winter month is a very real one. People who have been paying $30 a month in the winter to heat their homes with natural gas may well be paying $50 to $60 only a few winters hence. And the price of a gallon of gasoline is already shocking everyone who remembers the days when the order "fill 'er up" left change from a $5 bill. Sharply rising prices for the necessities of electricity, heat and gasoline will hurt most American families on a tight budget. Real hardships will likely be felt by fixed income and low income citizens unless they are given some form of relief.

Energy price increases are bad news for consumers but there is a silver lining in this cloud. As consumers begin to pay higher prices for energy in the 1970's, there will be a break with the wasteful consumption patterns of the past. The "more is better" philosophy, which brings penalties in the form of runaway power and fuel bills, will be replaced by a policy of thrift in the use of energy. The new era of scarcity and rising energy prices forces the nation to reexamine the vitality of competitive forces in the energy industry. The question of whether price control actually discourages industry from developing new supplies must also be debated and resolved. The fundamental question persists: Who is really responsible for the U.S. fuel supply? The government? The energy companies? No one? To answer that question and fashion the right mix of incentives and controls to assure adequate, reasonably priced supply is the task we face in the 1970's.

CHAPTER 8

Conflicting Maze of Government Policies

Energy shortages prompt us to demand to know who is responsible. Ordinarily, in a competitive market, a shortage is quickly remedied by entrepreneurs scrambling to fill the gap. But as we saw in the preceding chapter, competitive forces are weakening in the energy industries. These industries do not operate according to the classic model of the marketplace, for the government has intervened over the years to create a conflicting maze of programs that simultaneously regulate, promote, protect, finance, subsidize, compete with, and supply the energy companies.

The responsibility for meeting the nation's energy needs is divided between the companies in the business and the government. Unfortunately the division is so unclear that in fact no one is really responsible. Energy is sold by private companies, (except for publicly owned electric and gas companies) but government policy, and not the marketplace, goes far toward determining the amount and type of fuel available, the price charged and even the size of the market.

The federal government has erected many road signs to guide the companies, but the signs point in all directions and companies are understandably confused. As a company executive once remarked, "Industry can accommodate to any damn-fool policy of government, but it can't follow a question mark."[161]

In a very basic way, the federal government must shoulder the responsibility for future energy supply. Energy is not like oranges, which may be short one year but plentiful the next. Shortages will persist and can be cured only by expensive long-term research and development programs to enlarge the sources of supply. Thus if we foresee an acute energy shortage in 1985, investments in new sources must be made today —tomorrow will be too late. But companies can hardly be expected to invest hundreds of millions of dollars in new technology over a 10-15 year period with no payout until the end of the program and a high risk of no payout at all. The growing shortage of clean energy is thus probably beyond the capability of private enterprise to resolve. And with no national policy focused on achieving results in the time available, the energy supply problem has become worse each year.

Responsibility for solving the long-term energy problem thus falls heavily on government. But it is meanwhile the job of the private companies to keep the lights on and the family car running. As we observed earlier, this planet has not yet run out of energy resources in the ground. The companies may well have reasons of their own for allowing fuel supply to tighten. But it is difficult to detect their failures when they are able to hide behind the crazy-quilt of federal policies which make it difficult for the companies to perform their job.

Government energy policies have emerged piecemeal over the years as issues were tackled one at a time. As a result the government has instigated different regulatory and promotional programs for coal, atomic energy, natural gas, oil, and shale oil without recognizing that federal boosting of one source might discourage industry from investing in the others. Each source of energy is subject to a separate decision-making process, often by a different federal agency. An even greater problem is the impact of environmental protection, price control, foreign policy and tax policy on energy supply. New policies are made and new programs launched in these crucial areas without serious consideration to their effects on the demand for or supply of energy.

No effort has been made to examine the many pieces of the federal energy puzzle to see if they fit together in a way that fulfills the basic goals of public policy. Programs that subsidize and promote greater energy consumption persist in an era of energy scarcity. The income tax laws provide incentives to encourage the search for petroleum while price controls discourage the search. Environmental protection programs point one way and energy programs plunge ahead in the opposite direction blind to the conflicts. Atomic energy research is funded massively and all other options are starved for money.

There is no focal point for energy policy in the federal government with sufficient authority, staff and budget resources to reexamine the policies and reshape them into a pattern that can meet our urgent problems.

Viewed separately, many of the government programs in the energy sector of economy have been for the public good, at least as this was perceived at the time each new program was launched. Some programs were designed to benefit industry. But even with the benefit of hindsight few would question that federal programs such as rural electrification, basic utility regulation, and development of atomic power have generally served the people well. It is only when we assess these programs in the light of current and prospective resource scarcity, environmental concerns, and other emerging problems that they appear visibly inadequate.

The deep involvement of government in the affairs of the energy industries is by no means unnatural. It is necessary if the public interest is to be served.

Public utility regulation is one such intervention. The companies that sell electricity and natural gas to the consumer operate natural monopolies. It is costly enough to build one set of power lines to deliver electricity or a single natural gas distribution system to serve customers in a given locality. Duplicate service to individual customers for natural gas or electric service is out of the question with current technology. It would add 50 percent to consumer prices and would be extremely wasteful of the limited space available both above and below ground in metropolitan centers. Competition is not possible; at the same time, a consumer owning a gas furnace or an all-electric home can't buy energy from someone else. He is a captive customer and government price controls are essential to replace competition in assuring him a fair price.

Local utility regulation by franchise was the pattern of the 1890-1910 period, but it soon proved to be a failure.[162] The franchise was a contract, and the fixed conditions of price and service were much too rigid to regulate the rapidly growing utilities. Electric companies were expanding and interconnecting their local systems to achieve the economies of scale. A regulatory body with authority to make periodic reviews of the companies' operations as a whole was an obvious necessity.

State regulatory commissions were established beginning in 1907 with specific authority to control the rates and service of electric power and gas utilities (in those days it was gas manufactured from coal). By 1930 this regulatory pattern was established in most states. It remains the basic forum for controlling the retail price of electric power and natural gas as well as establishing the conditions of service under which utilities operate.

The state commissions set utility prices essentially on the basis of a cost-plus formula. A utility is reimbursed for all of its costs of service and allowed a reasonable profit. The margin of profit is based on a percentage of the net investment in plant and equipment (or its estimated value) and is not related to efficiency in keeping costs under control. Future rates were fixed on the basis of past performance. The regulatory commissions largely accept the costs which the utility has incurred.

Utility profits thus grow only when there is expansion of plant and equipment. Rates are shaped to stimulate growth through a sliding scale that offers reductions in price as each customer's volume of consumption increases. These rates are called promotional rates, and that is exactly what they are. With a pricing formula that ties profits not to efficiency but to growth, it is no wonder that utilities have devised pricing schedules that entice consumers to use energy lavishly. This may have made sense in the past when costs were declining, but now the faster a utility grows, the faster rates rise. The new era of shortages will require an altogether new approach to utility regulation, and a new emphasis on price to encourage frugality.

State regulatory sanction for promotional pricing by utilities heeded a national policy laid down by the federal government. Important new federal energy programs enacted by the Congress in the 1930's and 1940's had as a common purpose to "promote and encourage the fullest possible use of electric light and power . . . at the lowest possible rates."[163] These words are from the Tennessee Valley Authority Act, which established a government corporation that developed into a utility with responsibility for supplying power over an 80,000-square-mile region. TVA provided a promotional yardstick which Congress established to assist the regulatory commissions in measuring the performance of the private utilities.

The same promotional policy and philosophy was embodied in the laws that established the Rural Electrification Administration to provide low interest government loans to bring electric service to rural America, the Bonneville Power Act that required that electricity generated at federal dams in the Pacific Northwest be sold at rates "established with a view to encouraging the widest possible diversified use of electric energy,"[164] and the statute governing disposition of electricity of hydroelectric projects of the Army Corps of Engineers in various parts of the nation, which specified that the Secretary of Interior should sell the electricity "in such manner as to encourage the most widespread use thereof at the lowest possible rates to consumers consistent with sound business principles . . ."[165]

Looking to the future, there is reason to doubt whether either the existing regulatory system or the structure of the electric power industry can adequately respond to the new set of public priorities the nation's circumstances now demand. The regulatory system has painted itself into a corner. It still encourages rapid growth and the pollution that it brings at a time of scarcity when the need is to emphasize conservation and efficiency. At the same time the utility commissions are still judged by their ability to keep prices under control in an era when utility rates must be increased because of runaway costs the commissions cannot control. In fact, the very price fixing system it administers condones high costs by awarding profits on every dollar spent no matter how much a plant costs. Trapped by a system that has created public expectations of low rates, the commissions may not be able to raise rates sufficiently to cover the rapidly increasing utility costs. Both consumer and investor confidence in the utilities and the commissions that regulate them are very much in danger.

Utility rate regulation has also played a role in the utilities' neglect of research and development (R & D) for new technology. Such efforts in the past have been practically nonexistent. R & D expenditures by the utilities over the years have averaged only one-fourth of 1 percent of gross revenues as contrasted to manufacturing industries generally, which spend an average of about 2 percent. The electric power industry is now taking steps to enlarge its research efforts, but the modest goal is two-thirds of 1 percent. The industry is still faced with a regulatory system where the payoff is in keeping rates as low as possible to the consumer and where even if the R & D is successful, profit margins remain the same.

The structure of the electric power industry also has built-in obstacles to meeting the public's needs in the decades ahead. There is a serious concern whether the 3,600 municipal, cooperative, state, federal and private entities that now make up the electric power business can cooperate to build the kind of systems the nation needs in the future.

Structural flaws are nothing new. The task of rationalizing the electric power industry was begun in 1935 when Congress passed the Public Utility Holding Company Act in response to the scandals uncovered by the Federal Trade

Commission. The culprits were holding companies that owned the utility operating companies. The holding companies could have integrated the planning of their subsidiary utilities over a broad regional area, but they were more interested in bilking the operating companies of their profits and milking investors through the sale of watered stock. One device was to pay exorbitant fees to service companies which rendered little or no service. Unscrupulous promoters nearly brought the electric power industry to financial ruin.

The Holding Company Act, which resulted from the drive for reform, is administered by the Securities Exchange Commission. Within a few years, the SEC succeeded in separating the ownership of most utilities from the paper holding companies which owned them. Just as important, the SEC restored investor confidence in the industry. But financial reform was not all that Congress contemplated. The next step was to determine the size of utility companies that would best serve the public interest and to make recommendations concerning needed reorganization of the industry structure.[166] Almost four decades later, that Congressional mandate has yet to be carried out. It has now become an urgent matter.

Thus far the many small companies still in the generation and transmission business have not submerged their parochial interests. Most "power pools" consist of committees established to coordinate the expansion plans of the individual companies. That is a far cry from a corporate entity with authority to locate the plants on the basis of a region-wide perspective. The problem is exaggerated by antagonisms between the publicly owned and privately owned systems, many of which would prefer to continue yesterday's ideological battles rather than join in building integrated regional systems responsive to today's needs. New utility organizations are needed that are specially designed to satisfy the new multiplicity of public concerns, a subject we will address in Chapter 13.

The utilities have always lived in a halfway house between private enterprise and government domination. In contrast the fuel producing end of the energy business has been the epitome of the free enterprise system. Drilling for oil is a prime example of a risky operation in which an investor can strike it rich or go broke. Entrepreneurs attracted to exploration for petroleum would hardly be interested in an industry with a fixed profit and no hope of making an extra buck. It has thus been a rude awakening and a frustrating experience for petroleum producers in recent years to find their financial fate determined more by the policy of governments, which they cannot control, than by their own skill or luck in finding oil and gas.

Complaints of oil producers about government policy are ironic because over the years when government intervened in their affairs, it was to help them and not to get in their way. Government began its friendly intervention in petroleum affairs even before the enactment of the depletion allowance in 1926. Not as well known but even more important is the privilege—known as "expensing"—of counting all of the non-salvageable investments in exploration and development of petroleum as an expense in the year incurred. Each dollar so spent is an income tax deduction—instant depreciation so to speak, even though the well may produce for 20 years. Together these two tax privileges have given the petroleum industry a distinct advantage in attracting capital and have enabled companies in the oil business to pay relatively low federal income taxes. In 1971 the major petroleum companies paid less than 7 percent of their worldwide revenues in federal income taxes, as compared to a U.S. corporate tax rate of 50 percent. These tax breaks have survived intact over the decades except for a modest reduction in the depletion allowance from 27.5 percent to 22 percent in 1970.

The purpose of these special tax incentives is to stimulate exploration and development for petroleum, and in retrospect no one can seriously claim that they have been unsuccessful in encouraging more drilling. Investments in drilling have certainly been larger than would otherwise have been the case because the federal taxpayer was sharing the risk in a very large way. There is serious argument, however, as to how much additional oil is produced because of the tax breaks. Critics contend that with government underwriting most of the costs of a dry hole, the tax breaks encourage inefficiency and waste in drilling programs.

Perhaps the least defensible aspect of the tax breaks is their application to oil produced in foreign nations. It is at least arguable that the nation has a stake in special inducements to encourage development of domestic oil to avoid foreign policy and balance of payment problems with imports. This argument becomes ridiculous when we realize that foreign production also generates enormous tax relief for the American companies. The depletion allowance is applicable, but it is not needed because the Treasury permits the companies to count their large payments to foreign governments for oil production as a direct dollar-for-dollar offset (limited only by the proportion of revenue earned in each country) against U.S. income taxes. These payments are the largest item of cost in producing oil abroad and thus generate billions of dollars of tax deductions each year.

Foreign taxes that are in fact income taxes on company profits may be a legitimate deduction against U.S. income taxes. But the oil industry payments to the governments of the Middle East are not a tax on profits; they are the major item of cost in producing oil. These payments now have jumped from about $1.00 a barrel in 1971 to over $7.00 a barrel in 1973. They are obviously a cost of doing business and are in effect a royalty payment to the owner of the resource and not a tax on profits. Most Americans would not find it amusing if they knew that on January 1, 1974, when the producing nations added about $5.00 a barrel to the royalty payment which the oil companies must pay that these dollars indirectly have their origin in the U.S. Treasury. The depletion allowance and this technique of offsetting taxes are encouragements to the petroleum industry to drill in the wrong place at the expense of the U.S. taxpayers.

The ability to avoid a fair share of the federal tax burden has enabled the petroleum industry to price its products lower than would otherwise be the case and thus has encouraged more oil consumption. Here again we see a policy designed to promote exploration and increased use of energy persisting in an era of scarcity and concern for pollution, problems which become worse because of the greater consumption the tax subsidy encourages.

Perhaps the most distorting and unintended effect of the tax loopholes afforded the petroleum industry is that they greatly handicap the economies of alternative sources of energy which do not enjoy the same benefits. Shale oil and synthetic oil and gas made from coal suffer a handicap that may now approach $1.00 per barrel because of the more favorable application of the tax laws to natural petroleum. These are differences large enough to tilt investments one way or the other.

It is anomalous that the tax laws provide a big incentive to exploration and development for oil in Saudi Arabia and Libya while seriously handicapping shale oil development in Colorado—and all in the name of providing this nation with secure sources of oil! No one really doubts that the nation needs all the domestic petroleum that can be produced in environmentally acceptable areas. But the tax provisions cost the U.S. Treasury some $1.6 billion annually in potential federal revenue. It is time we stopped spending these huge sums in such a confused, ineffective, and counter-productive way.

The special tax provisions for oil producers also have a disturbing anti-competitive effect on the industry. The tax writeoffs apply only to the production of oil. An independent refiner or distributor of fuel oil or gasoline pays taxes like an ordinary businessman. However, the major oil companies produce, refine, and market their own products. Their tax deduction on production is large enough to reduce income taxes for their total operations to less than 7 percent on the average. Thus the tax breaks effectively reduce the cost of gasoline and fuel oil for the big companies which produce their own crude. This tax system makes it very difficult for a small independent to enter and compete in the refining and product distribution end of the business because the independent does not enjoy these tax benefits that are available only to a company engaged in production. An independent pays 50 percent taxes while the major integrated company pays 7 percent. In addition, an independent refiner must actually pay the prevailing price of domestic crude, which is merely a number on the books for an integrated company that produces its own crude.

The major companies set the price of crude and have every incentive to set it relatively high because the depletion allowance is 22 percent of the price thus set. A higher crude price for a company producing most of its own crude means a larger tax deduction and thus lower taxes on the business as a whole and a greater competitive advantage over independent refiners, gasoline stations and fuel oil dealers. It is more than a coincidence that 90 percent of the refining capacity in the United States is controlled by the major oil companies, which also control more than 80 percent of the retail gasoline sales.

Another set of policies favorable to the oil industry is production control or "pro-rationing." Production controls were administered by the conservation commissions in the producing states, but the program had the sanction of federal law. The Connally Hot Oil Act makes it a federal offense to ship any oil in interstate commerce that is produced in violation of state law. Offshore production on federal land areas was controlled in a manner parallel to states. Without this federal sanction it is quite likely that the state prorationing orders would run afoul of the anti-trust laws.

This particular "helping hand" for the petroleum industry assured that production exactly matched the demand for oil each month to avoid the prospect of price competition in years past when there was surplus producing capacity. It is still on the books, but not operative because most U.S. oil wells are now operating at 100 percent of maximum efficient capacity. The demand for petroleum appears quite likely to outpace the industry's ability to increase its productive capacity in the United States. Thus production controls are anachronistic. The program is worth recalling, however, as another example of government intervention to support the petroleum industry that has persisted for more than three decades. And it is still in "ready reserve" available to prevent price competition if domestic supplies should ever again exceed the market for petroleum.

In the late 1950's a major breach opened in the anticompetitive wall that the petroleum industry had erected with the support of federal tax breaks and production controls. Oil from the Middle East and Venezuela began to be imported in significant quantities, offering strong price competition to domestic producers. The federal government again came to the rescue in 1959 with a program that imposed strict quotas limiting the volumes of foreign oil that could be imported. The limitation was 12.2 percent of domestic production, thus guaranteeing the domestic producers some 88 percent of the market protected from price competition. During the 1960's consumers were deprived of billions of dollars each year in savings which the lower-priced imported oil could have afforded them.

The oil import quota system was justified in the name of national security, a vague and elastic phrase, but most students of the quota program outside the oil industry believe its major function was to protect the domestic industry against foreign competition. It is perhaps the most glaring example of federal intervention in the marketplace to prevent competition in the energy industries.

Debate over the oil import control program continued throughout its 13-year history, but the industry was politically stronger than its critics. The program finally outlived its usefulness, even for the industry. By early 1973 the price of oil in the world market had approached domestic prices. More importantly, supply was so short that domestic production was not threatened. Indeed it was clear that the U.S. would suffer severe shortages unless imports were permitted to flow in whatever quantities the marketplace demanded. The President's Energy Message of April 18, 1973, finally buried the import quota program.

Thus we see that over the years the federal government has intervened again and again to favor the oil industry and protect it from competition. It was not until the environmental revolution that the government clashed directly with the interests of the oil producers and began to reflect the broader interests of the public.

The federal government has not, however, been quite so

generous to natural gas, even though it is a by-product of oil in many wells and is produced largely by the same companies. Natural gas enjoys the same tax privileges as oil production, but the big involvement of the federal government in natural gas production began in 1954, when the gas producers got the unhappy news from the Supreme Court, deciding the Phillips Petroleum Corporation case,[167] that they were subject to federal price controls. This intriguing chapter in the involvement of government in the energy industry has dominated natural gas production in subsequent years.

The Natural Gas Act was vague with respect to FPC authority over sales at the wellhead by the producer to the pipeline. Congress had passed an amendment to the Natural Gas Act in 1948 to remove the doubt by exempting sales by natural gas producers, but President Truman vetoed the bill on the grounds that such regulation might be needed at a later date to protect consumers from windfall profits by the producers. It was not until the Supreme Court's decision in the Phillips case that it became clear such sales were also subject to Federal Power Commission price control. After the Phillips decision, Congress again passed legislation that would exempt gas producers from FPC price control in 1956. This legislation was supported by the Eisenhower Administration but the President vetoed the bill because a scandal occurred as it moved toward passage (see Chapter 9).

Price controls over natural gas producers remain intact but they are still a matter of serious dispute and efforts continue to overthrow them. This particular government intervention has definitely been on the side of the consumer but interestingly it was triggered by a Supreme Court decision and barely escaped a Congressional death sentence.

The Mandate of the Natural Gas Act for natural gas producers and interstate pipelines alike is to fix prices at the "lowest reasonable rates."[168] It should be noted that this statuary mandate was in keeping with the prevailing philosophy of promoting the greatest possible consumption of energy at the lowest possible price, which was embodied in the earlier statutes controlling electric power during its rapid growth as a source of energy.

The Federal Power Commission succeeded in keeping natural gas prices at levels far below the price of alternative fuels with a consequent savings to consumers of billions of dollars over the last two decades.[169] Until very recently natural gas was found largely as a byproduct of the search of oil. It was, therefore, low in cost and seemingly abundant. The Federal Power Commission had a mandate to assure that the consumers and not the producers received the benefits of this low cost source of energy, and to a large extent the FPC succeeded. We will discuss the future role of price controls over natural gas as part of an overall national policy in Chapter 13.

As imports of liquefied natural gas from the Middle East and the Soviet Union became a significant element of energy supply, a new government intervention enters the picture. Imports of liquefied natural gas are being heavily subsidized by the federal government. The government pays about half the cost of the LNG tankers built in the United States and through the Export-Import Bank the American taxpayer is assisting in the financing of liquefaction plants built abroad. For example, an Algerian project recently approved by the FPC will cost the federal government $134 million for 6 of the fleet of 9 tankers. In addition, the Maritime Administration guarantees 75 percent of the ship mortgage.

The debate over federal price controls on natural gas has merged into the broader question of price controls in the economy generally as crude oil and coal have come under federal price controls. No one can be sure of the future course of government anti-inflation policy. But for more than two years this form of governmental intervention has been having a considerable influence on how much money is invested in fuel industry expansions. Coming at a time when fuels are already in tight supply, the controls point up the dilemma the public faces. The energy industry expands only if the price is right, and it is the judge. Thus the public is left with the Hobson's choice of prices the government deems are too high or shortages, or both. Surely federal policy must come up with a better solution for the public than that.

The coal industry has remained closer to the free enterprise pattern than perhaps any other segment of the energy-producing industries. There was a brief and abortive attempt at federal support for coal prices in the 1930's.[170] But by and large the coal industry has been competitive and the prices of its products have been relatively free of economic regulation by government. But coal has recently been presented with a federally created competitor as well as a host of federal environmental health and safety and price control programs that have all but eliminated its ability to grow, until it can accommodate to these concerns of society.

Another major governmental energy program has created a powerful competitor for coal. The Atomic Energy Act was passed by Congress in the aftermath of the use of the atomic bomb to end World War II. This act, passed in 1946, created the Atomic Energy Commission with a mandate to develop atomic energy for peaceful purposes. The AEC has been successful and the research effort continues with federal investments on the order of $700 million dollars annually to devise more efficient nuclear power plants.

The government's single-minded devotion to atomic energy research over the years has stacked the deck for the future in favor of the one energy source it has so lavishly supported. It is not wrong for the federal government to fund the development of atomic energy. But it was wrong and terribly distorting to fail to mount a comparable program for coal and other attractive energy options. The coal industry can hardly expect the government to pass up the atomic option and sit idly by while the energy crisis worsens. But the public is not well served when government chooses only one corner of a broad field, especially one in which private companies must compete with no comparable research support.

The lesson here is that the federal government did not "think energy" but rather pursued the potential of the atom oblivious to other opportunities or the impact that atomic power development would have on competing forms of energy. As a consequence the nation faces the future with coal —our most abundant source of fossil fuel—largely unusable for lack of research in mining and conversion technology. Industrial incentives for doing the research were nonexistent in the face of competition from a federally financed alternative source.

The uranium industry actually began with only one customer—the government. It was not until 1970 that any private company could own uranium ore or enriched fuel. Today uranium is sold in the open market, but the price is still more influenced by the federal government than by the law of supply and demand. The market price for uranium is supported by a cozy combination of government programs. The government maintains an absolute ban on the importation of uranium for domestic use, even though lower-cost ore is available on the world market. There is no pretense that the ban is needed for national security. It is carried out pursuant to a statutory mandate to preserve a viable domestic industry.[171]

In addition, the government owns a stockpile of uranium valued at $800 million which is surplus to its own needs. This stockpile is being disposed of on a very slow timetable over the next decade in order to avoid having any appreciable impact on the market price of uranium. This program, too, is justified on the theory of protecting an "infant" industry until it can become viable. But this infant industry is no babe in the woods. It is made up primarily of the major oil companies who now own 50% of United States uranium reserves.

The future plans of the uranium industry are very much controlled by federal policy. There is the ever-present possibility of permitting imports or deciding on a more rapid disposition by its surplus stockpile. Either action could greatly influence uranium prices. Perhaps even more crucial is the continuing federal monopoly on the facilities to enrich uranium ore to a grade suitable for use in power plants. Thus the key to the availability of nuclear fuel is still in the hands of the federal government.

Thus far we have reviewed federal programs for economic regulation, subsidy, protection, and ownership of energy facilities. The producers of energy and the consuming public are now dependent on government in a much more fundamental way. The government owns most of the remaining oil and gas, coal, uranium, shale oil, and other fuel resources. The most prolific areas for petroleum are off shore in the outer continental shelf beyond the state boundaries where the resources are owned by all the people of the United States. Most of the remaining coal and uranium as well as shale oil and geothermal energy is located in the western lands that are also federal property. Thus an energy producer must do business with Uncle Sam if he wants to stay in business. And in so doing a new uncertainty is injected in his ability to meet future energy needs.

The problem is that the government is in the very awkward position of really not knowing much about the resources it owns. In the past, government has relied heavily on industry to identify the areas that it should offer for sale and has been quite responsive to industry. But the recent concern for protecting the environment is changing the rules.

The revolution in public concern for protecting the land, oceans, and beaches has centered on opposition to federal leasing programs. The Santa Barbara oil spill in 1969 was a direct result of a policy which placed more emphasis on oil exploitation and the financial returns to the Treasury than on protection of the environment. And before 1969 over 10 billion tons of coal in the West were leased by the Interior Department with little or no enforcement of requirements that the land be reclaimed, or even knowledge of whether reclamation is possible.

The National Environmental Policy Act now requires that the federal government look at the environmental consequences before leasing its land. But the government has yet to fashion a leasing program that will accommodate the public's concern for environmental protection and its equally strong interest in fuel supply. The Interior Department's record of catering to industry means that environmentalists will continue pushing for legislation and court orders to turn off the federal "energy spigot." As a result, the basic energy supply for the nation is put in jeopardy by an energy-oriented federal department's failure to respond to the environmental ethic and the new laws' requirements.

The government sale of leases to produce fuel is another of the programs dealing rather directly with energy industries. The influence, distortions, and uncertainties of these programs persist into the future. Yet, in a sense, industry has learned to live with government energy programs. What caught the industry totally unprepared was the new wave of government intervention in recent years through laws, regulations, and standards to restore the environment.

The environmental and health and safety laws enacted in the last five years have superimposed a whole new set of federal edicts on top of the existing interventions. The environmental laws have established new goals. But the real challenge lies ahead in developing programs to implement those goals, and to do so requires rationalizing the new environmental ethic with the energy consumption ethic that still permeates the land.

The air pollution control laws alone are having a revolutionary impact on energy supply and demand. As far as demand is concerned, control standards are tilting the market toward natural gas because for many uses it is the only source clean enough to meet control standards. Unfortunately natural gas is in shortest supply. The next best source of energy to satisfy air quality standards is oil, especially low sulphur oil. But new supplies of low sulphur oil are also hard to come by. Air pollution controls have thus exacerbated the gas shortage and also shifted the market toward oil, which is scarce and must be imported. And to complete the picture, the standards effectively outlaw the use of the domestic fuel which is most abundant, namely our enormous supplies of high sulphur coal.

The air pollution laws are but the most important of the multitude of new programs for environmental protection that tend to short circuit the ongoing system of energy supply. Most of the environmental problems discussed earlier, have given rise to governmental programs in an attempt to cure them. The programs that have affected the energy industry include the water quality laws, bans on strip-mining, regulations limiting offshore drilling, health and safety laws for coal and uranium miners, land-use restrictions on sites for power plants and refineries, and environmental impact statements for all federal actions. Each of them has established a whole new set of ground rules to which the energy companies are being forced to adjust. The answer to the environmental crises is not to create an energy crisis, but neither is the answer to turn our back on the environment. All the pieces of the policy puzzle must be fitted together in a manner that recognizes both the need for energy and the need for protecting the environment.

The obvious response is the development of a national energy policy that recognizes the concern for protecting the environment. A first step in developing such a policy is to recognize that it takes time to turn around a society. Environmental protection must be coordinated with the available technology of the energy industry. Otherwise government will turn out the lights in a vain attempt to clean up the air. But even more important is the need for new policy directions for the energy agencies and the industry. Both of them are moving sluggishly and defensively in response to the environmental ethic. To move full speed ahead in restoring the environment, we must devise more precise marching orders, orders that can in fact be carried out.

The government has intervened in the energy market over the years with an almost unbelievable array of programs. In the early years the thrust was for greater exploitation, subsidies to develop new sources and government regulatory programs to keep prices low for consumers. More recently it has intervened to protect the environment, and this has made the impact of the prior interventions seem slight by comparison. When the new programs to clean up the environment were added to the old program to encourage and promote energy, the result was certain to be a conflict. Consumer protection has taken a back seat, and the drive for protecting the environment has given way to the fear of shortages. The energy companies quite naturally are dragging their feet, reluctant to change direction, and seem content to take advantage of the government's confusion.

CHAPTER 9

The Politics of Energy

The petroleum industry, and to a lesser extent the utilities, virtually dictated U.S. energy policy in the past quarter century. The politics of energy have been dominated by the politics of Texas and the Southwest, whose oil producers made the most of their unity of purpose and abundance of cash. They were also fortunate to have such spokesmen as Sam Rayburn, Lyndon Johnson, and Robert Kerr who ran the Congress and exerted powerful leverage to make certain that national energy policies were acceptable to the domestic producers. Their task was made easier by the almost total lack of interest and concern of the rest of the country.

But the politics of energy are now undergoing revolutionary change. Rising concern over pollution in the late 1960's and more recent shortages and soaring prices, have made the energy companies villains in many people's eyes. Energy policy has moved off the financial pages and onto the front page of the nation's press. The country at large has become interested, indeed angry. And as energy policy proves to be an issue that swings votes from New York to California, the people's interests begin to shape policy.

Even so, energy laws and programs still look as though they were largely made in Houston, rather than Washington, D.C. It is no accident. The industry has spent large sums of money on advertisements, political contributions and lobbying to get its point of view across. And in the past they have been most successful. The petroleum industry in particular makes large contributions to Presidential campaigns—contributions which assure their representatives of access to high places and a strong voice in the appointment of key government officials. In the 1972 campaign, the oil industry contributed an estimated $5 million to President Nixon's re-election alone.[172]

In contrast to the intense interest of the energy companies, consumers until recently have tended to ignore energy policy questions. Prices were low, supplies abundant and other issues more pressing. Over the years the energy companies had the federal government and most state legislatures pretty much to themselves.

For example, during the 1960's few citizens even knew about the oil import quota system, much less realized that it was adding two to three cents per gallon to the price of their gasoline. Those who were aware were too few to make a difference politically. The same was true of the depletion allowance and other oil industry tax loopholes, which survived "reform" with only minor changes. The oil industry had the good fortune, foresight and influence to assure that the congressional committees handling tax matters were dominated by sympathetic congressmen and senators. As for agencies less friendly to the petroleum industry and the utilities, such as the regulatory Federal Power Commission, the industry was able to influence the appointment of commissioners over the years so as to make regulation more an illusion than a reality except for occasional bursts of reform activity.

More than public indifference was necessary for the oil industry to achieve an energy policy of promotion, depletion, and protection. The strategic leadership role in the Congress fell to the delegations from such producing states as Texas and Oklahoma. But the most important factor was the petroleum industry's very effective lobbying. Using money, lots of money, a swarm of industry lobbyists worked day in and day out to sell their story.

The Directory of Petroleum Industry Washington Representatives lists 60 different oil and natural gas company organizations with people in Washington, D.C., to represent their interests—about one organization for each federal agency that deals with petroleum. These 60 organizations include trade associations such as the American Petroleum Institute with a budget of $17 million (but reporting only some $40,000 in lobbying expenditures in 1972). Apart from the trade associations, almost every major oil company has a Washington representative. There are 425 full-time employees on oil industry payrolls in the capital, and a conservative estimate of $25,000 per year for each employee, including overhead, would put the petroleum industry payroll for visible Washington representation at more than $10 million per year. No one can estimate the additional sums spent in less visible ways, for expense accounts and the like. Moreover, the payroll figure does not include the fees paid the lawyers who advise the companies, represent them in proceedings before regulatory agencies and lobby for and against particular pieces of legislation. Suggestive of the scale of these expenditures, but perhaps not typical, are the legal fees and expenses that one company, El Paso Natural Gas Company, reported for 1972, almost $2 million.

The petroleum industry undoubtedly has the largest lobbying contingent in Washington (almost as large as the Congress), but the other sectors of the energy industry are by no means under-represented. The electric power industry, the coal industry and the utility equipment manufacturers all have their men in Washington, and they do not lack for funds.

The public is seldom even aware of the intensive lobbying conducted by the energy industries. Occasionally the industry has overreached and been caught in tactics which were so heavy-handed that the public took notice. One notable example was in the effort in 1956 to repeal the Federal Power Commission's price controls on natural gas at the wellhead. The repeal bill was rolling smoothly through the

Congress under the guidance of two Texans, Senator Lyndon Johnson and Representative Sam Rayburn, the respective majority leaders in the Senate and the House. But an oil industry lobbyist happened to leave twenty-three $100 bills in a brown envelope in the office of Senator Francis Case of South Dakota. Senator Case angrily disclosed this apparent bribery attempt on the floor of the Senate. The incident caused a public sensation but did not alarm the senators already committed to vote for the decontrol bill, because the bill was passed. But President Eisenhower, despite his earlier support for the legislation, then vetoed the bill because of the "arrogant" tactics used by the industry, which he described as "in defiance of acceptable standards of propriety."

The apparent attempt to bribe Senator Case represents a rare instance in which someone erred so badly that the conscience of the nation was briefly troubled. It was an instance where a "fourth rate oil lobbyist,"[173] as Lyndon Johnson reportedly referred to him, tried to use the power of the purse in such a blatant way that it back-fired. But what is significant is that the oil purse is powerful, it is used, and it is effective. Campaign contributions running into the tens of millions of dollars are intended to influence the government in the same way as those twenty-three $100 bills left on Senator Case's desk. The difference, unfortunately, is that campaign contributions by the special interests meet "acceptable standards of propriety."

Actually, well-heeled oil lobbyists merely provide the point of contact for the industry. The real source of influence in Washington is the industry's influence back home. The depletion allowance and oil import quotas were issues that swung elections in Texas and other producing states. A Presidential candidate's decision to oppose the depletion allowance could cost him Texas while losing few if any votes in the disinterested nation at large. If, in addition, large campaign contributions also were at stake, it is little wonder that Presidents have routinely acquiesced in the crucial demands of the oil industry.

For decades both political parties have thought of energy policy mostly in terms of capturing the electoral votes in

Texas. For example, over a 13-year period, the import quota system to protect domestic oil producers was continued primarily because of the White House's keen interest in carrying Texas and not antagonizing the powerful leaders in Congress who represented the oil industry's interests.

Oil issues swing votes in Texas, Oklahoma and other producing states because a large number of influential citizens have a direct, personal monetary stake. Not only the major oil companies but also thousands of small independent producers of oil and gas pack the political wallop. Independent oil producers tend to be local notables with standing in their communities. They influence a wide circle of people. A sympathetic press and the state's own stake in production (the Texas schools are largely financed from oil revenues) combine to make the depletion allowance as politically potent in Texas as civil rights laws are in Harlem.

In every state and most congressional districts, the electric power utility and the natural gas company wield substantial influence, and in many eastern states, the coal industry has been strong. The leaders in these energy industries tend to be prominent citizens in their own right. And when energy issues do not concern the public at large it is relatively easy for these executives to use their influence with senators, congressmen and White House aides.

The energy industry influence on the Congress as a whole, of course, does not stem from anyone's interest in "carrying Texas." But congressional leaders from the oil producing states have controlled these levers of power for the better part of two decades, especially during the heyday of that mighty triumvirate—Speaker Rayburn, Senate Majority Leader Johnson, and Senator Robert Kerr, the Oklahoma oil millionaire. The absence of opposition over the years enabled senators and representatives from the petroleum producing states to take advantage of the senority system in Congress and gain positions of vast power and influence.

In more recent years, too, senators and congressmen from the oil states have occupied key positions. Take, for example, the 92nd Congress (1971-72): In the Senate, Allen Ellender of Louisiana was Chairman of the Appropriations

Committee, a position of immense leverage, and his colleague from Louisiana, Russell Long, stood guard over the tax laws as Chairman of the Senate Finance Committee. In the House, Carl Albert of Oklahoma was Speaker and Hale Boggs of Louisiana, the Majority Leader. The influential House Appropriations Committee was headed by a Texan, George H. Mahon, and the Armed Services Committee was chaired by F. Edward Hebert of Louisiana. The deaths of Senator Ellender and Representative Boggs thinned the leadership ranks of men from the oil states in the Congress in 1974, but others continued in influential roles.

Actually, the political bloc of "energy states" is growing. As we saw in Chapter 7, the oil companies have been buying into coal and becoming diversified energy companies. A community of interest has thus developed between oil and coal that could create a powerful West Virginia-Texas axis in the Congress. The West Virginia congressional delegation includes Senate Majoriy Whip Richard Byrd and House Commerce Committee Chairman Harley Staggers. In fact, if coal states such as Pennsylvania, Ohio, Illinois, and the Appalachian and Rocky Mountain states were to align themselves with the oil states they could easily dominate the Congress on energy issues.

But new forces are at work in the Congress which threaten energy industry domination as never before. Congressmen from throughout the country are hearing from their constituents on energy questions. Energy policy is fast becoming a major consumer-oriented issue for members of Congress, and that includes the producing states as well. People are suddenly interested, and this new reality of public interest and wide publicity diminishes the influence of the oil industry and the energy companies.

At the White House and within the executive branch of federal government the energy industry has long exerted its influence through campaign contributions and by manipulating every administration's desire to "carry Texas." The energy establishment seeks at least veto power over the nomination of federal officials to key policy-making positions and hopes to influence their decisions. One means of exerting influence is simply the day-to-day contact between government energy-policy officials and industry spokesmen. The industry also makes effective use of the leverage that key congressmen sympathetic to the industry can exert on executive branch policy. Regulatory officials and other Presidential appointees know that their chances of reappointment and retention in office can be jeopardized by making "trouble" for the industry. They know how politicians in the White House tend to react to the combination of industry influence and consumer apathy.

Presidents have often used nominations to the Federal Power Commission and proposed changes in the oil import quota program as trading points with legislators who have constituents or supporters to whom these are vitally important questions. An FPC seat has been used more than once to gain congressional support on broader and more important issues. For example, when the late Senator Everett M. Dirksen of Illinois was Senate Minority leader, he bargained for his utility and natural gas pipeline "clients." He was "persuaded" to support U.S. financing of the United Nations in 1963 after Harold Woodward of Illinois was appointed to the FPC, thus assuring the industry of a commissioner who voted a straight natural gas industry ticket. President John F. Kennedy obtained the backing of Senator Kerr of Oklahoma, then Chairman of the Finance Committee, for the Trade Expansion Act of 1962 only after he agreed to continue the oil import quota program.

The energy industry takes an intense interest in the appointment of FPC officials who fix the price of natural gas and electricity, the AEC commissioners who decide whether a nuclear power plant is safe, and the Interior Department officials who enforce the coal mine safety laws and fix the terms by which the government's petroleum and coal is to be sold. In the past the public hardly knew these officials existed. The point is often made that energy policy officials should have industry experience in order to know what they are doing. But this argument misses an essential point. What these officials are doing is making public policy, and for that purpose they need an appreciation and concern for values cherished by all segments of the public. To be sure, there is a need for reliable technical expertise at the staff level of the energy policy agencies. But the policy-makers should at least begin their terms in office with a perspective and viewpoint that is oriented toward public concerns. They will have time enough to learn the industry point of view.

Once in office, energy policy officials are subject to subtle but effective influence through the trade press. This influence is exerted on behalf of the industry's policy line. Only a person who has worked closely with policy-making officials can fully appreciate the impact of the daily trade press treatment. The *Oil Daily* and *Platt's Oilgram* bring the energy establishment the word from Washington every day, and they are supplemented by the *Oil & Gas Journal, Electrical World, Nucleonics Weekly, Nuclear Industry* and other publications that make up the reading material of federal officials. In the past the general press was as disinterested as its unaware readers. Everyone quite humanly likes to read favorable things about himself in print. Energy policy officials receive blanket coverage in the trade press; a consumer-oriented energy policy official takes a daily beating in the only press that covers him regularly. He quickly acquires the lonely feeling of a soldier who turned left when the rest of the troops turned right. He soon begins to feel that all those consumers out there whom he is trying to protect could not care less about his good intentions. Sadly, this impression has not been altogether inaccurate in the past.

The Department of the Interior is a good example of an agency dominated by the energy industry insofar as policy positions are concerned. Few people in this country had heard of Steve Wakefield, who in 1973 was Assistant Secretary of the Interior for Mineral Resources, or of his predecessors during the past decade, Hollis Dole and Cordell Moore. But the oil industry knew them well before they were appointed to this critically important office. It is no coincidence that both Dole and Moore went to work for the oil industry after leaving office.

Throughout the 1960's the Atomic Energy Commission was a fraternity of the atomic energy establishment virtually closed to outsiders. The AEC Chairman, Glenn Seaborg, had discovered plutonium; Commissioner James Ramey had served as staff director of the Congressional Joint Committee on Atomic Energy; and Commissioner Wilfred Johnson had directed an AEC-funded laboratory in the Pacific Northwest. For "balance" there was Commissioner Clarence Larson who had managed an AEC installation in Oak Ridge, Tennessee. These men were able, honest and dedicated public servants who acted out of conviction. But they started with an unquestioning faith in the overriding necessity to develop the atomic power industry and a tendency to ignore problem areas. The result was an agency that coddled the industry and stressed promotion of the atom—their lifework—over the kind of objective, tough-minded approach to safety and other regulatory programs that an independent agency such as the Environmental Protection Agency has shown in enforcing air pollution control laws. The pattern has been that the Atomic Energy Commission, the congressional joint committee on atomic energy, and the nuclear industry worked together in harmony and with little effort to involve the general public, believing that the subject matter of their shared concern was too complicated to involve outsiders.

The cozy relationship of the AEC and industry is typified by an episode in 1973 when a contract was signed between a utility and a manufacturer to build an atomic power plant in the Atlantic Ocean off the coast of New Jersey. The idea of building a nuclear power plant in the Atlantic Ocean was an altogether novel one, to say the least, and it required an AEC license, which could be issued only after a formal hearing on the issue of whether such a plant was safe and compatible with the marine environment. The project was launched with some fanfare at a signing ceremony aboard a yacht near the proposed site. Among those in attendance were James Ramey and Clarence Larson, two of the AEC commissioners who might later sit in judgment on the license application. It was as though an FPC official had stood next to a utility company president at a press conference announcing an application for a rate increase. Yet for the AEC commissioners the event was routine. Indeed, the AEC's intimate relationship with industry is so commonplace that the incident was hardly noticed in the press.

Presidential appointments to the Federal Power Commission in past decades have followed, with some notable exceptions, a slavishly pro-industry pattern. Public interest in the energy field was at such a low ebb in the 1950's that the President was not required even to pay lip service to consumer protection. When the term of FPC Commissioner William Connole expired in 1960, President Eisenhower did not reappoint him and publicly gave as a reason that he felt there should not be a consumers' man on the FPC. Presidents subsequently were more subtle, and from time to time consumer advocates were even appointed. But they found the going tough indeed as pressure was applied through the Congress and the White House.

White House pressure on regulatory officials is seldom overt; it's simply that consumer-protection men are rarely reappointed. It is no accident that three recent consumer-oriented chairmen of the Federal Power Commission—Leland Olds, Joseph Swidler, and Lee White—each served only one term. Olds' reappointment was defeated in the Senate in 1949 by opponents led by Senator Lyndon Johnson. Joseph Swidler, who was appointed by President Kennedy, was not renominated by President Johnson. Instead Johnson appointed another consumer man, Lee White, his White House Counsel, who had not yet incurred the wrath of industry. President Nixon quickly replaced White with Chairman John Nassikas, who is behaving more in keeping with the long-term industry domination of the regulatory agencies.

When consumer-oriented energy policy officials launch actions that offend the powerful, the pressure can be terrific. For example, in the early 1960's when FPC Chairman Swidler began implementing price controls over natural gas producers and electrical utilities, he incurred the wrath of Congressman Albert Thomas of Houston, Texas, whose House Appropriations Subcommittee oversees the FPC's budget. A few months before President Kennedy's death, Chairman Swidler found himself summoned to the White

House and informed by Larry O'Brien, the President's chief lobbyist, that Swidler had troubles on the Hill and that Congressman Thomas was a loyal administration supporter on many key issues. Several months later, with a Texan in the White House, Chairman Swidler appeared before Congressman Thomas's subcommittee seeking funds for his agency at closed hearings. The following exchange from the transcript eventually made public, provides only a very limited glimpse of the vicious personal attack Swidler endured:

> MR. THOMAS. You are bragging on yourself a little too much; are you not? You never saw a gas well, did you, before you got this job? Your experience has been in the electric power industry and TVA. What do you know about oil and gas? Are you trying to tell us now all the brains in the gas industry are over in the Federal Power Commission?
>
> MR. SWIDLER. No, sir.
>
> MR. THOMAS. You did not say it in so many words, but you have intimated that. You have done this, you have done that, you have done the other. It never had been done before. Who furnished all these brains? What did you know about the gas industry before you got this job? Did you ever see a gas well before?
>
> MR. SWIDLER. No, sir.
>
> MR. THOMAS. You learned quickly; did you not?
>
> MR. SWIDLER. I am just trying to do as good a job as I can, sir. I make no claim to being an expert, and in my confirmation hearings I said very frankly that I knew very little about the oil and gas business.
>
> MR. THOMAS. You have done the impossible, starting from scratch. So, ipso facto, you are the greatest gasman in the world and know more than all of them put together. This was in 3-1/2 short years.
>
> MR. SWIDLER. Two and a half, sir, as Chairman. . . .
>
> * * *
>
> MR. THOMAS. What is sacred about Federal regulation? We have it thrown in our faces every day that the Federal

Government is moving into industry, taking over. Washington is taking over. Here you are leading the procession. You are hard to carry come November, you know that.

MR. SWIDLER. Mr. Chairman, the question whether there should be a Federal Power Act is not my question. That is a congressional question.

MR. THOMAS. You make it your question when you want to enlarge your jurisdiction. You are saying regulation has just gone to pot. You are the savior riding the great white horse, going to save the country. We are asking you: Where are the wrongs? Who has commissioned you to cure the wrongs and what are those wrongs?

MR. SWIDLER. I am trying to answer your question, sir. I start by saying we did not write the act. We are only commissioned to administer it. The act provides for wholesale rate regulation. It is not for us to repeal it, sir. It is for Congress to repeal it if it wishes to. It is for us to administer it. . . .

MR. THOMAS. May I interrupt? As to your jurisdiction—I speak for myself and I imagine for most of the members here—we all are for regulation, we know the value of it. I, for one, just do not want you to overregulate anybody.

MR. SWIDLER. No, sir. I agree.

MR. THOMAS. We have that thrown at us, we on this side of the table. We have to go to the people every 2 years. You never have to go to them. We have to carry you.

In 1965 President Lyndon Johnson left another FPC Commissioner, Charles R. Ross, hanging in limbo for more than a year after his term expired, because of industry opposition. This opposition was reinforced by Johnson's belief that Ross should never have been appointed in the first place. Apparently President Kennedy had promised to check out his FPC seat with Senator Kerr of Oklahoma as part of the price for going along with the appointment of pro-consumer Swidler. And President Kennedy had failed to do so. Finally Ross was reappointed, but only after another commissioner had died and President Johnson was able to couple Ross's appointment with that of a protégé of Senator

Everett Dirksen, Carl Bagge, who was expected to look after industry interests with the not inconsiderable support of his sponsor, the Senate Minority Leader.

But for once the system broke down. Commissioner Bagge, the Dirksen protégé, was a last-minute choice because other possible nominees were too blatantly pro-utility for Johnson to accept. Bagge, a railroad lawyer, appeared reliable enough to industry, but to everyone's surprise he turned out to be a consumers' man on many crucial votes. Indeed he was soon considered a traitor by petroleum industry and natural-gas-pipeline supporters. He even voted against the FPC's endorsement of one of Senator Dirksen's own legislative proposals designed to give the utilities a special tax break. As a result, few people gave Bagge much chance of being reappointed when his term expired in 1967.

But again the system failed. As fate would have it, former Counsel to the President, Lee White, was chairman of the FPC, and he was a man trying desperately to hold on to his pro-consumer majority. Using his knowledge of White House procedures, he arranged for Commissioner Bagge's reappointment to be sent up to the Congress routinely by the President many months before his term expired. When Senator Dirksen heard the news of Bagge's reappointment he stormed over to the White House to protest to President Johnson. The President, with a look of shock and consternation, said: "Why Everett, if I had only known that you didn't want your own man reappointed I would never have sent his name up. It never even occurred to me that you were opposed to your own man." Senator Dirksen realized that he had been had. It was impolitic for him to complain that he was against his own man simply because that man had been honest and had voted his own convictions. President Johnson reportedly enjoyed his one-upmanship of Dirksen immensely. But the story does not end there. After Senator Dirksen died several years later, Carl Bagge resigned from the FPC to return to private life as the president of the National Coal Association, and is belatedly making good on the Senator's judgment that he would look out for the energy industry's best interests.

The energy companies had a long winning streak going through Democratic and Republican administrations, and they had every reason to expect to continue to win with the election of President Nixon in 1968. Although the consumer movement had gained strength, energy had not yet become a swing-vote issue in the nation at large. In the late 1960's, only New England and the Southwest exhibited a political interest in energy policy. The reason was simple: Consumers in New England pay the highest price for energy in the country. Even in New England, however, energy issues were only one of a myriad of problems plaguing the consumer. The attention that consumers could give to the subject was slight in comparison to the central role that petroleum plays in the economic life and politics of Texas and Louisiana.

Under Nixon, the Federal Power Commission and the Department of Interior remained securely in industry-oriented hands. Interior Secretary Rogers C. B. Morton revealed the Department's attitude in a White House briefing of petroleum industry officials on August 16, 1973:

> In conclusion, let me say that the Department through its Office of Oil and Gas, through the Office of the Secretariat, in the whole energy and minerals area, Steve Wakefield has the responsibility, and is staffed up to handle that responsibility. Our mission is to serve you, not regulate you. We try to avoid it. I have tried to avoid regulation to the degree that I possibly can. We want to be sure that we come up with guidelines and programs where guidelines are necessary, that have a maximum of input by people who make their living in the marketplace. I pledge to you that the Department is at your service. We cannot be all things to all people. We cannot straddle issues. We have to do business today and tomorrow.

For a brief interlude in 1972, James Schlesinger, an outsider to the energy establishment with a zeal for the public interest, headed the Atomic Energy Commission, but he was called away to become Secretary of Defense before his efforts to reform the AEC into a full-fledged regulatory agency could bear fruit. The Federal Power Commission's current Chairman John Nassikas and his fellow commissioners have identified the FPC with the industries' objectives of higher prices and resistance to strict enforcement of environmental laws. All in all the energy companies managed to continue to dominate the federal energy establishment.

Yet, in the early 1970's, they found themselves in an entirely new ball-game with the government—one in which they were losing badly. Despite campaign contributions and influence-seeking, the political climate for the energy industries suddenly turned stormy. The turn-around can be summed-up in one word—environment.

America's awakening to what energy production and consumption were doing to the air and the earth came about suddenly as television dramatized the activities of the energy industry and brought oil spills and air pollution into the living rooms of American homes. Environmental protection became a "motherhood" issue, and suddenly the politics of energy took a dramatic turn for the worse. The energy industries came into direct conflict with the health concerns and the environmental ethic of most Americans. And the energy industries discovered that when the public becomes aroused about an issue, special-interest money and influence are of very little help because politicians know the public and the press are watching them. When the public concentrates on an issue, the public usually wins, for few politicians want to run against the crowd. The special interests win out in the dark but usually lose in broad daylight.

This is not to suggest that the environmental movement consists only of "little old ladies in tennis shoes" writing indignant letters to congressmen. There are some two dozen environmental protection organizations with offices in Washington, D.C. Some are activist lobbying groups with very small staffs (usually 3 or 4 persons), such as The Environmental Policy Center and Environmental Action. In contrast, the National Wildlife Federation has a membership of 2.5 million and a Washington staff of 50 but does no lobbying. Actually, there are only about a dozen people in the activist groups committed to open lobbying on environmental issues. The other Washington-based environmental groups rely on the strength of their national membership and grassroots pressure to influence government. They are anxious to protect their status as organizations eligible to receive tax-deductible contributions, a privilege which would be jeopardized if they engaged in open lobbying.

It is ironic that the energy industry engages in lobbying and deducts it as a legitimate business expense, with the taxpayer paying the bill, but that environmental or consumer groups seeking to establish a lobbying organization cannot obtain an income tax deduction for the donors of the money they spend. Even with such one-sided rules, the environmental protection forces have skillfully brought their broad public support to bear on government officials, aided by prominent citizens and reinforced by the mass-media's attention to the issues.

The energy industries have tried to create an anti-environmental backlash by suggesting that environmental protection laws should be rolled back to permit energy production to proceed. The industry has conducted a massive and expensive advertising campaign. In 1971, the American Petroleum Institute began the effort with a $4 million television and newspaper campaign built around the slogan: "A country that runs on oil can't afford to run short."[174] The campaign was supplemented by a $2.5 million opinion-molding effort aimed at community and business leaders. The natural gas and electric power industries sponsored similar advertisements. Individual energy companies such as the Mobil Oil Corporation and Continental Oil Company ran individual campaigns which included full-page ads in major newspapers and radio commercials throughout the country. Such industry advertising is doubtless having an impact on the American public. As shortages of fuel and gasoline became more prevalent, the ads appeared to make a valid point.

The energy industries are counting heavily on the fear of shortages—the energy crisis—to turn the tide in some of the battles they have been losing to the environmentalists. The federal government holds the key because most of the fuel resources to be exploited are located on public lands in the west or offshore. And it is the federal government that sets the pace in air pollution control and licensing of atomic energy. An "energy crisis," properly managed and manipulated, could be a godsend for industry to override the environmentalist opposition. The energy crisis can also serve as a reason, or excuse, to exempt petroleum and coal from price controls. Shortages can reasonably be said to result from "unwise and counter-productive" price controls that have inhibited the search for more fuel.

But it is by no means clear that the public will accept the industry's version of the energy crisis. Another public reaction to shortages and higher prices is that a conspiracy is afoot. To many consumers the so-called energy crisis is a fake, manipulated by the companies in league with the government to rip off the consumer. To these consumers the industry advertising campaign and hard-sell simply reinforce their suspicions.

The energy companies are also faced with two dynamic factors that are changing the political scene. The Watergate scandal has undermined public confidence in the credibility of every large institution. The fact that an $89,000 contribution to President Nixon's 1972 campaign by Texas oil man Robert H. Allen, President of Gulf Resources and Chemical Company, was "laundered" through a Mexican bank and used to help finance the illegal activities only dramatizes the connection. All this comes, moreover, at a time when consumers are steadily gaining strength in their lobbying tactics. "Participatory democracy" is attracting the time and attention of talented people, and energy problems are now touching people generally. The energy companies can no longer win policy battles by public indifference and default.

A little-noticed example of the changing climate was the fate of President Nixon's latest industry-oriented appointment to the Federal Power Commission—Robert H. Morris, a San Francisco lawyer who for years had represented Standard Oil Co. of California before the FPC. On June 13, 1973, for the first time in the 53-year history of that agency, the Senate rejected a candidate for the FPC because of his industry background. As Senator Warren G. Magnuson, Chairman of the Commerce Committee said: "Now, more than ever, the Senate should not be asked to confirm appointments to regulatory agencies which appear to have been designed as rewards for politically supportive industries or other special interest groups."

The rejection of an industry-oriented FPC nominee is but the latest evidence of the erosion of industry influence since the "good old days" when Sam Rayburn and Lyndon Johnson ran the Congress and the public was uninterested in the politics of energy. The depletion allowance, the symbol of oil's political power, actually was reduced in 1970 after surviving intact for 44 years. The reduction of the depletion allowance from 27.5 to 22 percent was limited in its impact because other provisions made the effective rate less than 27.5 percent for many companies in any event. Also the more important loophole—"expensing of intangibles"—remained unchanged. Nevertheless the handwriting had appeared on the wall.

The energy industry's campaign to use the public's fear of shortages as a lever to regain the policy initiative is by no means a guaranteed success. Many politicians will not embrace the industry line. Making the environmentalists appear to be the villains who caused the energy crisis clashes with the basic values of a large part of the population. The public knows it is the energy industries that are polluting the air and that have been unwilling or unable to take actions to comply with the laws protecting the environment. The shortages we have experienced are a measure of the industries' failure as well as the failure of the federal government to anticipate the need for research and development and other programs to assure adequate supplies of clean energy.

The shortages of the past two years coupled with concern for protecting the environment have moved energy policy into one of the most crucial battlegrounds of national politics in this decade. Not only does the issue test the depth of our dedication to preserving the environment; more important, it also challenges the growth patterns and lifestyles of America's future. Energy is at the cutting edge of the rising debate over whether more is always better. Shortages can sway people to believe that industry is right, that we must dig faster and pay more for energy; but a sense of crisis can also lead to basic changes in our assumptions and our values. Most Americans, and even the industry interest groups involved, have yet to appreciate fully the dimensions and the critical nature of the political struggle beneath the energy crisis.

In the past, the politics of energy have been dominated by regional interests and have not become an issue between the two major parties. By and large, the Democrats have been as good to the petroleum industry as the Republicans. But the new battle over energy policy will be central to the future of America and people will choose sides. Inevitably, the major parties will follow suit.

Five to ten years ago anyone in public life opposing projects to exploit our resources or build new industrial developments would be committing political suicide. At that time, anyone proposing to exploit the coal resources in Montana and other western states would have been greeted with open arms as a boon to employment and regional economic growth. The same was true in West Virginia and other Appalachian states, where increased coal production was regarded as the remedy for the poverty afflicting the region. But these attitudes are now undergoing revolutionary change. Widespread public realization of the ravaging effects on the landscape of strip mining and other shortsighted industrial development is turning opinion around.

People, of course, are still interested in full employment, but those in the sparsely populated states are shifting their priorities. The commercial interests favoring exploitation of fuels and development of power plants and industry are now at least matched by people favoring slower and more balanced growth. A new value is emerging in our society. More and more people are becoming dissatisfied with material affluence alone and are opposed to the head-long rush toward industrialization. They see more damage than benefits to the community and their own lives from new industry, new developments and more people. This trend is apparent in the Pacific Northwest, New England, Appalachia, and the Southeast, and the suburban areas of every state.

The rancher in Montana, the fisherman in the Pacific

Northwest, the mountaineer in West Virginia, the tenant farmer in South Carolina, the hiker in New England, and the cross-country skier in Minnesota—all these and like-minded citizens may have nothing more in common than a love of the earth the way it is. There may even be a strong streak of individual selfishness in this new movement, a feeling that we have made it and we don't want other people crowding in on us. One might sarcastically define an environmentalist as the fellow who bought his ski lodge last year. Whatever the motive, the movement to slow growth is a very real and increasingly powerful political force.

The opposition of local citizens to power plants, oil refineries, strip mining and like projects stems in large part because while most of the mess is left for them, the benefits go largely to people in cities far away. It is difficult to persuade the people of Colorado to ravage their countryside in order to supply fuel for Chicago and St. Louis and other population centers. True, we are one nation. And consumers in New York are being fueled by natural gas from Louisiana. But in former days, the oil and gas producing states were eager to sell because they received large financial benefit and perceived no burden. Now the producing areas see their countryside and lifestyles endangered to satisfy wasteful consumption in faraway cities. The proposals for strip mining of coal, oil shale development, and construction of large power plants are now the subject of sharp disputes at the grassroots level. The energy companies face serious opposition in what was formerly docile territory. At the moment, the Chamber of Commerce argument—we need the jobs—may still be dominant, but public opinion is shifting rapidly to a new emphasis on slower and saner growth.

Most of the people in this country live in metropolitan centers and they have yet to choose sides in the political fray over energy policy. The energy industries and the current Republican administration no doubt regard the business community and middle-class America as a core group of supporters. They stress the discomforts that energy shortages bring and the need for increased energy supplies to keep factories humming and jobs plentiful. Their policy stresses rapid development of America's energy sources so we can get back to business as usual and be independent of any foreign government's influence. The strategy is to join labor, much of the Jewish vote, and middle-class America, and to drive a wedge between these political-economic interest groups and those moved by environmental concerns.

Suburban America represents the middle ground where the issue of energy policy is likely to be decided. Suburban Americans are energy guzzlers who lead lives dominated by their automobiles, their large homes and the mechanical gadgets they contain. They are certainly worried about fuel shortages. But suburbanites also know that much else is worrying them. They may be virtual captives of their present patterns of energy consumption, but they are also frustrated by traffic jams, cars and appliances that break down frequently, and the encroachment of high-rises and other urban developments that destroy the very breathing space for which they moved out of the city in the first place.

Working men and suburbanites might both choose the industry side of the energy battle but chances are that most of them will not. The people who live in the cities breathe the air made foul by the massive consumption of energy. They know that air pollution control is not a luxury but a vital safeguard to their health. City dwellers are not likely to be impressed with efforts to discredit the clean-up programs. And suburbanites who are concerned about the threats of rezoning and unplanned growth are likely to remain open-minded on ways to solve the shortages by means that imply slower and more efficient growth. If they don't, their children may see that they do.

Perhaps more important for the consumer than these considerations is the fact that they pay the bills for all the energy that is consumed in the U.S. In the final analysis, the pocketbook issue probably will be decisive for city dwellers, suburbanites and Americans generally. The slogan "Save Energy—Save Money" may have more appeal than one would have thought a year ago when we were still enjoying the afterglow of an era of low-priced energy. Consumers are already discovering that less energy consumption saves them money and also hastens the cleaning up process of the air they breathe. And for the Jewish voters and citizens generally who are concerned about the leverage on U.S. foreign policy from imports of Arab oil, an all-out policy of saving energy may well seem more attractive than one of "Dig We Must." As energy bills sky-rocket, shortages may cause consumers to turn against the companies which raise prices and then fail to meet their needs. In a society becoming more cynical and suspicious, the hard-selling of the energy crisis could itself provoke a backlash.

As the problems of energy are better understood, as the environmental movement realizes energy policy is the key to environmental clean-up, as consumers begin to recognize how much money is at stake, and as we begin to better understand the cause of shortages and the options available for avoiding them, it is entirely possible that energy policy could become as much a people-oriented issue as the environment. Energy policy was scarcely mentioned in the 1972 presidential campaign. It is bound to be a major issue in 1976. By then both parties will be vying for the consumer's side of the energy issue.

PART III

PATHWAYS TO SOLUTION

CHAPTER 10

Opportunities for Energy Conservation

The government-industrial energy complex has told us repeatedly that overall energy consumption in the United States must double by 1990, and that electric power consumption must continue its growth pattern of doubling every decade. Otherwise the poor will remain poor, the rich will have to give up their comforts, and a great many people in between will be thrown out of work. Energy consumption is used as a yardstick to measure our standard of living, and even the quality of life in America. A message has been drilled into our heads for decades: "More is always better."

Most Americans therefore were shocked as they began to learn that energy consumption in the U.S. is as much a measure of our society's waste and inefficiency as it is of useful work performed.[175] Waste is apparent throughout the energy supply system.

Only one-third of the oil found in a petroleum reservoir is lifted out. Only 20 percent of this oil's energy potential is put to useful work in the ordinary automobile engine. Thus only about 6 percent of the energy potential in an oil reservoir is actually put to work in powering the family car. The other

94 percent is either left in the ground or wasted out the exhaust pipe of the car. We leave half of the fuel in the ground when we deep-mine coal. Of the coal that is mined, more than 60 percent of the energy potential is wasted in conversion into electricity. Thus if the electricity is generated from coal mined underground, only about 15 percent of the energy potential in the earth actually reaches the consumer. Even then, after the various forms of energy finally reach our homes, offices, automobiles and industries, a substantial fraction of the remaining potential is lost because in the past no one thought of designing anything with an eye toward conserving energy. On the contrary energy seemed as plentiful and cheap, and we used it lavishly.

The nation has been blind to the growing waste of energy. This was a natural consequence of an era in which energy sources were abundant and prices low. In many instances it was even economically attractive for us to use energy freely and wastefully. Subsidies and promotional pricing of energy made it economical to use more rather than invest in equipment that would use less.

The fact that we waste more than half of our potential energy is important news, and in a sense good news. We can eliminate waste and thereby avoid potential shortages. It is already technically feasible to eliminate much of the waste. What is even more encouraging, rising prices have made it economically feasible to save more and more energy.

It is obvious that material affluence and energy consumption are related. Energy is required to heat our homes and run our dishwashers, cars, and factories. The key fact is that we can sustain our standard of living with less energy than we are currently using. We can learn to get more driving miles out of a gallon of gasoline, and we can prevent heated and cooled air from leaking out of our houses. Our material comforts can continue to improve while the growth rate of our consumption of energy tapers off. We can mine coal, keep our homes cool, get to work, or manufacture a washing machine in a variety of ways, and we have abundant opportunities to choose ways that will substantially reduce our total requirements for energy.

To be sure, some growth in energy consumption in the next decade is a necessity, but how much growth is required and for how long are still open questions. The answers may spell the difference between worsening energy shortages and adequate supplies. Whether we can avoid shortages in the next 20 years will depend to a large extent on the efficiency with which we use the energy resources now available to us. The only long-term solution to our basic problem, of course, is an enlarged supply of cleaner energy through new technology. But a policy of saving energy can buy the time needed to perfect the new technology and put it into production.

It is also important to seek opportunities to shift consumption from scarce to relatively abundant energy sources. If, for example, the projected demand for oil actually does require massive imports that will create difficult foreign policy and balance of payments problems, then we obviously should determine whether our society's needs can be met with other sources which create fewer problems. Electric mass-transit and rapid inter-city railways would not only conserve energy but also shift our dependence for supply to more plentiful domestic sources of electric power.

Savings are possible at every link in the energy supply chain. The following chapter examines opportunities for greater efficiency in the extraction of fuels from the earth and their conversion to a useable form, processes that are under the control of the energy industries. In this chapter, we will look at energy use in the home, the automobile, and the factory—the places where energy is ultimately consumed, and where major savings will require knowledge and action by society as a whole.

It will be instructive to examine each of these areas with an eye toward identifying how we can eliminate energy waste. We will also be alert to what we may lose or gain through these savings. Is it really possible to improve the quality of life while slowing our rate of growth in energy consumption?

In buying a home, an air conditioner or until recently even an automobile, consumers were concerned only with the purchase price or, typically, the size of the down payment required. The fuel and power bills that would follow were commonly ignored. Now that shortages have pinched the consumer and prices are rising sharply there is new consumer interest in yearly operating costs.

President Lyndon Johnson was ahead of his time when he kept turning off the light switches at the White House, practicing energy conservation at the point of consumption. There is a large multiplier effect that ripples back through the entire energy supply system when a consumer saves energy in his home, car, or factory. For example, just by purchasing a more efficient unit, a homeowner can air condition his residence with 25 kilowatt hours a day rather than 50—a quite feasible saving—and in doing so save the equivalent of 75 kilowatt hours of fuel. This is true because it requires 3 units of fuel to generate one unit of electricity. The two units of fuel wasted at the power plant are saved as well as the one unit of electricity saved in the home. Although the multiplier effect is not as great in other forms of energy the consumer uses, the inefficiencies in every energy supply system magnify the opportunity for savings at the point of consumption.

During the past year Americans have become energy-saving conscious. We have been forced to a number of rationing measures to overcome a sudden shortage. Some of these measures, such as more moderate indoor temperatures, should become a permanent feature of our lifestyle. Others such as rationing are obviously undesirable. Our problems this year are really the end result of trying to satisfy the appetite of a nation of energy gluttons. The nation needs to build into its growth policy an energy-saving ethic. We shall be discussing here the more basic actions needed to squeeze the waste out of our energy system so that we can indeed balance our energy budget in the years ahead.

ENERGY SAVINGS AT HOME

Until early in this century, people seeking protection against the elements relied primarily on the building in which they lived or worked. Their expectations were not great. A single fireplace and a stove provided warm spots around which the family gathered. For the rest, they depended on the building, thick walls, with few windows or other openings, to keep out the winter cold and summer heat. Firewood has long since given way to low-cost, abundant supplies of more concentrated forms of fuel. And the American people have steadily increased their expectation for comfort indoors. Central heating is now a commonplace feature of American life and most other nations in northern climates are rapidly following suit. Air conditioning is following the pattern of heating.

Americans expect—indeed they insist upon—sufficient energy to maintain the temperature they desire in every inch of indoor living and working space no matter how hot or cold it may be outdoors. Space conditioning has reached the point that by most sensible standards the interior of the typical American building is too hot in the winter and too cold in the summer. Larger and larger quantities of fuel are required to achieve these desired but unnecessary temperatures.

Fuel requirements have grown larger as builders have relied less and less on the buildings themselves to insulate the space indoors. The buildings of a century ago with thick masonry walls and heavy slate roofs have been replaced by buildings with large expanses of single-pane glass and roofs that are thinly insulated. As a result, enormous volumes of basic fuel and electricity are required to make the indoor environment more and more "comfortable." From 1960 to 1970 the energy consumption of the average U.S. household rose by one-third. Energy supplies have been so abundant and so low in price that it seemed unimportant that most of the energy purchased to heat and cool homes leaked into the outdoors.

Buildings in the U.S. on the average consume about 40 percent more fuel each year for heating and cooling than would be necessary if they were designed, insulated and built to use energy efficiently. Leaks are found in the same places in all buildings. The walls and roofs are inadequately insulated; wide expanses of glass permit heat loss; there are too many openings around doors, windows and chimneys. About twice as much heat and coolness escapes through the walls as would be the case with optimum insulation. An important savings can be made through a relatively small investment in thicker insulation, by a more judicious use of glass as a building material, and by operating heating and cooling systems more efficiently. These investments typically would pay for themselves in 3-5 years. Thereafter, the consumer would save energy and money too.

Space heating and cooling are by no means minor items of energy consumption. Space conditioning of residences accounts for 12 percent of total U.S. energy consumption; in commercial buildings, where similar savings are possible, it accounts for an additional 9 percent. If industrial buildings are added, we find that more than 25 percent of the total energy used in the U.S. is consumed in heating and cooling the buildings in which people live and work.

Making energy conservation an overriding concern in the design, construction and operation of new buildings could result in a savings of about 40 percent of the energy per cubic foot of space now being consumed. This would mean a reduction of about 8 percent (40 percent of 20 percent) in the total volume of additional energy required each year in the future.

But half of the buildings in the U.S. that will be standing in the year 2000 have already been built. After construction is completed, stopping the waste of energy presents practical problems. Most existing buildings are inadequately insulated, but homeowners and owners of commercial buildings are not likely to go to the expense and inconvenience of making such "repairs." Actually, added insulation for a roof is simple to install and inexpensive. And insulation can be blown into existing walls. But this sounds like a messy operation, and few people want to be disturbed in their homes unless they can be motivated by the fear of shortages or large monetary savings. Even so, lack of ready cash would stop most people.

Nevertheless, as we have seen in the past year, energy-saving measures are possible in existing buildings: Installing storm windows, plugging leaks around windows and doors, maintaining more moderate temperatures, turning down the heat at night, shutting off unused rooms, and turning off the air conditioner on cooler days. Even without heavier insulation, energy consumption for heating and cooling existing buildings can and is being reduced by about 20 percent, or half as much as in new buildings.

The challenge is how to translate the new interest in conserving energy in both old and new buildings into a program of actions that will capture most of the savings. As the price of energy increases, the incentives for action to save energy become even more attractive. The crux of the problem in the marketplace is that builders do not live in the structures they build. They sell or rent them, and it is the family or tenant who had no direct voice in the design who ultimately pays the bill for fuel and electricity. And the builders' incentives are to economize as much as possible on insulation, which isn't seen anyway.

Government action that requires builders to make economically feasible investments to save energy probably will be necessary. Building codes already contain insulation standards, and these need to be upgraded and enlarged to include all the features that could economically be incorporated in new and existing buildings to minimize energy consumption. The requirements for energy conservation would, of course, need to be continually reviewed and revised in light of higher fuel prices and new building techniques and materials. Such a system might borrow from past experience but apply it differently. For example, the utilities had no trouble developing standards for the "Gold Medallion" home when they were promoting electric heat. The state public service commissions could adopt rules that required new construction to meet demanding energy conservation standards before utility service would be made available.

Another approach would be to require that these energy conservation criteria be satisfied before the government would guarantee a loan by the Federal Housing Administration, Veterans Administration, Small Business Administration or any other agency for a new or existing home or commercial building. Such a requirement would go far toward getting the job done. On the average, homes change ownership every 5 years and rental property turns over every 2 years. It might be possible in a relatively short time to assure that most residences in America would be fully insulated either when built or when resold or rented. Still another approach would be to offer the homeowner an FHA loan to insulate his home or an incentive in the form of an income tax deduction for installing needed insulation. There is very little doubt that a serious federal commitment to stop the waste of energy from "leaks" in buildings could be successfully implemented.

It is not the building alone that wastes energy. Small changes in our living habits and attitudes are yielding large savings of energy and money. For example, those people who dialed down their thermostat from 72° back to 65° at night and 68° during the day this past winter found out they saved over 10 percent on their heating bills. One need not awaken in a cold house or even take the trouble to remember to turn back the thermostat at night; inexpensive clock thermostats can be installed that automatically adjust the temperature. Another simple but important step is to close drapes at night in the winter for added insulation and to open them during the day to take advantage of the sun's radiation.

Periodic inspection of furnaces to insure efficient operation can save 5, 10 or even 15 percent on fuel bills. Most gas furnaces have pilot lights that burn all summer and waste as much as 10 percent of the total energy consumed in a home. These pilot lights can be replaced inexpensively with an electronic starter. Fireplace dampers should be kept closed when not in use. Weather stripping windows and doors can reduce heating fuel waste by as much as 10 percent. And storm windows can make an equally significant difference.

Air conditioners vary by a factor of two in their efficiency, so that by spending only a little more on the unit itself, a consumer can save as much as half the energy required to maintain a desired temperature. Energy consumption for air-conditioning skyrockets without adequate ventilation. Ventilators and attic fans are easy to install both in new and existing homes.

The opportunities for important savings extend beyond the energy consumed in conditioning space within buildings. About 15 percent of the energy consumed in homes and apartments is used to heat water. The electricity needed to run a dishwasher or washing machine is trivial compared to the energy required to heat the water. Yet hot water is used only once and then the heat goes down the drain. Mechanical features could reuse this heat in the home, except in summer. An energy-conscious American will go easy on the hot shower and not run the hot water all the time he is shaving. In many households, the water heaters are too small and thus waste a great deal of energy in a futile effort to heat water faster than their design permits. Energy could be saved by using a larger heater or even two heaters. Water heaters are also often poorly located, and heat is wasted in delivering the water half way across the building in pipes that may not even be insulated.

New technology can help conserve energy in the home. In the next chapter we will discuss the solar house, which could use the rays of the sun for most of the outside energy needed for heating and cooling. A marvelous invention called the heat pump is already available. It pumps solar heat into a building from the outdoors even on cold days. Depending on the climate, a heat pump provides the consumer with 2 or 3 btu's of heat for each btu required to operate the pump. In the summer the machine acts as an air conditioner. In most regions of the nation, a heat pump can save two-thirds of the energy requirements in an all-electric home. And if the heat pump is powered by oil or natural gas, which is possible, it can cut in half the fuel required to heat a building.[176]

Heat pumps operate on the principle of refrigeration, and have been available for decades. Early models had a problem of reliability, but thousands of heat pumps now are providing satisfactory service. Until very recently, fuel prices were so low that there was little interest and perhaps scant economic incentive to making widespread use of the heat pump. The reason is that a heat pump adds to the capital expense of a building, costing some $1,500 for a 2-ton unit to serve a 1,600 square foot house. This is an extra investment because a furnace or electric resistance heating would still be needed in most parts of the U.S. for supplemental heat on very cold days. But with fuel costs at current high levels, the savings would pay for the heat pump in a few years.

In the past, energy companies wished to sell more of their product, not machines that would reduce their sales. But with fuel in short supply and prices soaring, public interest in conserving energy is mounting. The heat pump should become a best seller. But the necessary first step is for consumers to know of its availability and potential.

Savings can also be gained in changing the type of fuel used to heat and cool homes. In the absence of a heat pump, at least twice as much basic fuel is required to heat a comparably insulated building with electric heat than with natural gas or fuel oil. This is true because of the tremendous waste of fuel—some two-thirds—in its conversion to electricity at a large power plant. Even so, in the future, consumers may not regard electric heat as a costly mistake. Electricity is the only form in which atomic energy can be used, and thus electric heat will provide a new source of energy, rather than a wasteful use of fossil fuels, most of which are already in short supply. And equipped with a heat pump, an electrically-heated home is not an energy waster.

Today, the per-capita energy consumption in U.S. households is extremely wasteful in both economic and energy terms. Yet vast quantities of additional energy will be required if decent housing is to be made available for all Americans. Per-capita energy consumption must be substantially increased for the 25 percent of the American people who are poorly housed. And an almost unimaginable new energy demand will be created as impoverished people around the world begin to be adequately housed. But there is every reason for such new housing to be built in a way that reduces energy consumption 40 percent below the level in existing buildings. As noted above, much of the waste in these older buildings can be eliminated over a period of years. The savings realized can fuel the new housing required.

Within the current level of per-capita energy consumption in the home, every American can be provided with a comfortable indoor environment in summer and winter and an adequate array of appliances. To do so, however, we must eliminate the waste from our existing pattern of energy use. This will not require drastic changes in our life styles. It will require the right combination of consumer awareness, economic incentives, and government policies to make energy-saving profitable and popular (see Chapter 13).

COMMERCIAL ENERGY

About 15 percent of total U.S. energy is consumed in the supermarkets, tall office buildings, small shops, schools and other buildings where Americans work and shop. About half of the energy consumed in these commercial establishments is used for space heating, with air-conditioning, lighting, water heating, refrigeration and cooking the other significant end-uses. As a general rule a commercial building consumes more energy in proportion to its size than does a home because of the mechanical equipment, computers and other machinery it contains, and the need for greater ventilation.

The opportunities for saving energy are roughly the same in the commercial sector as in private residences and if anything, somewhat greater. It does not require an energy expert to see the waste in the typical office building with walls of glass windows that can't be opened, lights blazing like a Christmas tree and a heating and cooling plant that is usually operated around the clock even though the building is empty more than half the time.

Savings of 30 percent can be achieved by reducing the use of glass and installing double- or triple-pane glass in new buildings. Glass need not be eliminated. There are attractive new glass materials that darken when exposed to direct sunlight, and these minimize heat and cooling losses.

But a more fundamental change is needed—a change in the way architects and designers *think* about buildings. Over the past several decades they have given little or no thought to the orientation of buildings with the sun, or the shape of buildings that would conserve energy. The sheer vertical walls in new buildings afford no protection from the elements. These thin-shell, flat-roofed structures require enormous quantities of energy in order to maintain the year-round comfortable temperatures that are now standard in commercial establishments.

Architects and designers thus are presented with a major challenge: To design buildings that are pleasing to the eye and functional, yet more efficient in the use of energy. Pioneering efforts by individual architects are already showing the way. Innovative designs go beyond merely providing for adequate insulation and proper orientation to the sun. A new building in Arizona has walls at a 45° angle that act as sunshades.[177]

Most commercial establishments produce a sizable volume of solid wastes. These wastes can be used as fuel (see Chapter 11). A significant part of an office building's heat could be obtained by simply burning the trash in a power plant designed and built within the building. If buildings were well-served by public transportation, the space now devoted to underground parking garages could be used for on-site power plants, which would at once solve waste disposal problems and conserve fuel.

The greatest opportunity for energy conservation in commercial establishments is in improving the efficiency of their operation and maintenance. A tremendous volume of air is sucked into these buildings for ventilation which consumes 25-50 percent of the energy required for heating and cooling.[178] The intake of air into most buildings is greatly in excess of the volume required for a healthful indoor environment. Indeed, it is often excessive by a factor of four. Levels of ventilation needed in toilets and kitchens during the day are often used for entire buildings day and night. By dividing space into ventilation zones, air intake could be cut to a small fraction of current levels.

Savings can also be achieved by reusing energy rather than allowing it to escape in the exhaust air. The heat wheel is a mechanical device that keeps energy from escaping. It is placed in the main duct system and as it rotates it transfers energy from the exhaust air to the incoming air. The heat wheel is not a recent invention but its use is new. It can transfer about 80 percent of the energy to the incoming air stream and achieve savings of as much as 30 percent of all of the energy required for heating and cooling a commercial building. If the intake ducts are not located next to the exhausts, an alternative technology, called the heat pipe, can perform the same function, transferring the btu's of energy from the exhaust air to the incoming air.

In most buildings, no study has ever been made of ways to conserve energy. In fact the heat balance is so poor in many buildings that both heating and air-conditioning systems operate simultaneously on many days of the year. Significant savings can be made by paying more attention to the humidity level and by operating buildings at somewhat lower temperatures in winter and somewhat higher temperatures in summer than are currently maintained. Such minor changes would improve the level of comfort and save energy as well. Each building needs systematic evaluation—a heat balance. It should consider the heat that people give off, the light that comes in through the windows, the heat from lights, the fact that furnishings in a building store heat, and the kind of work being done. Heating and cooling requirements then should be tailored to the building's actual needs.

Heating and cooling equipment use energy most efficiently if operated close to maximum capacity. Fans and pumps consume 40 percent of the energy used by an air-conditioning system. But heating systems are usually designed to accommodate extreme weather conditions which may occur only once in several decades. Major savings could result from designs that would enable systems to operate closer to design capacity on most days of the year. Work itself can be redesigned. The four day week could have a favorable impact on energy consumption. If one assumes that most buildings are heated and cooled for the longer work day in any event, the shorter work week could cut down on commercial energy consumption by 20 percent.

Lighting in commercial establishments, including schools and office buildings, is at levels of intensity much brighter than necessary. Experts estimate that lighting levels could be cut in half. This is true even though florescent lights, which require less than half as much energy as incandescent bulbs per unit of light, are widely used. A chief reason why most buildings are grossly overlighted is that the standards for lighting intensity are set by a committee largely made up of representatives of General Electric, Westinghouse, and the electric utilities. The levels of illumination used in American office buildings exceed the levels used abroad by a factor of two. And this high level of intensity is not concentrated in the areas where it is needed, such as over a drafting table, but is used uniformly throughout every hallway and corner of a building. No adjustment is made for light available from outdoors. And, of course, the lights usually remain on for many hours after the work force leaves. Lighting accounts for as much as 5 percent of the energy consumed in commercial buildings. In the summer additional energy is required for air conditioning to combat the heat from the lights, which compounds the waste. The energy used in lighting is a glaring example of our energy waste. Savings on the order of 25 percent can be made immediately by turning off one-fourth of the lights in every building.

A final point about commercial buildings is more fundamental. There is large scope for savings of energy by use of more efficient building materials. The steel, reinforced concrete and structural materials used in most buildings are grossly in excess of what is needed with careful design and construction. The design codes for structural steel and reinforced concrete reflect an era when such materials and the large amounts of energy required to produce them were abundant. Safety factors are piled on safety factors to eliminate the need for careful design and to minimize labor. These codes urgently need to be reformed to reflect current and prospective shortages of resources.

Extra materials don't make buildings any safer. What is needed is a somewhat greater investment of man-hours in assuring the proper placement of steel and greater care in the pouring of the concrete in construction form-work. These are the areas where accidents now occur. By a somewhat larger investment of labor in these areas we can not only save lives but also as much as 50 percent of the material used in most buildings. These savings would significantly reduce energy consumption because the manufacture of cement and structural steel requires a great deal of energy.

The commercial building thus offers opportunities for saving energy and money through design and operation that adequately consider the economic balance between initial investment, comfort, and operating costs. In new buildings, savings of 40 percent in energy consumption are economically attainable. And in existing buildings, revamping operations and upgrading thermal insulation can achieve savings of at least half that magnitude.

TRANSPORTATION

So far as energy consumed for transportation is concerned, the distance that separates homes and commercial buildings is the crucial factor. Since World War II mushrooming suburban development has imposed an ever-greater need for the consumer to drive his automobile to work, to shop and to visit. Cities have assumed the shape of larger and larger "doughnuts," with the central city the hole in the middle. The elaborate circumferential highways encircling urban areas, and the traffic jams on them, illustrate how much most Americans must travel simply to get to work and back home each day. Eight out of 10 people commute by car because public transportation is unavailable or extremely inconvenient.

All forms of transportation account for 25 percent of total U.S. energy consumption. When the energy required to build the transport vehicles and the highways is added, transportation accounts for more than one-third of total U.S. energy consumption. The automobile alone consumes more energy than the heating and cooling of all U.S. residences. And its efficiency is even poorer. A 3,000 pound vehicle powered by an internal combustion engine uses gasoline at an efficiency of less than 20 percent to haul a 150-pound person. Unlike the heat that leaks through a building and merely wastes energy and money, the excessive volume of gasoline consumed ends up fouling the air we breathe.

The potential for energy saving in transportation is even greater than the 40 percent figure that can be achieved in the residential and commercial markets. Eliminating the waste does not require new inventions, but it will require basic reform of the nation's transportation system. These reforms are needed not only to avert an energy crisis but also to solve the problems of getting to work in a reasonable time and cleaning up the air in the urban centers.

The transportation market is a virtual monopoly of oil. Gasoline and jet fuels are the two petroleum products that power the automobiles, trucks and airplanes which transport about 85 percent of all passenger and freight traffic. In terms of energy consumption, the automobile is to transportation what space heating is to buildings. Automobiles consume about 55 percent of the total energy used in transportation, divided almost equally between urban and inter-city automobile travel. Trucks consume more than 20 percent of the total energy for transportation, and aircraft, the fastest growing category, are above 10 percent.

The automobile is almost a caricature of how Americans waste energy. The picture of the modern American being driven to work by a team of 300 horses is patently ridiculous but real. On the average, a gallon of gasoline is now required to drive a car 12 miles, a figure which has been declining steadily in recent years because of larger engines, accessories such as air conditioning, and new air pollution control devices. A requirement that all new automobiles be built to travel at least 24 miles on a gallon of gasoline, even allowing Detroit the time to retool, would cut energy consumption for automobile travel in half by 1985, as compared to today's mileage. This would mean less powerful engines, and lighter and smaller cars, but Detroit could still offer a family car and even a station wagon. Americans might be forced to slow down a bit, but they would lose little of their mobility.

The opportunities for saving energy in transportation go beyond requiring more efficient automobiles. In urban areas —where half the automobile miles are driven—mass transportation can save one-half of the energy that automobiles now consume. The need for improved mass transportation has been well documented. But what has been missing up to now is an appreciation of how important mass transportation is to solving the energy crisis. Mass transit has the potential for reducing the nation's largest single item of fuel consumption to less than one-half of current levels.

Electrified transit systems require many years to build, but a greatly expanded modern bus system, operating on express lanes of highways, could quickly become a reality if the U.S. were seriously concerned about averting the energy-environment crisis. There is much argument about the economics of public transportation. But mass transport is economical on any accounting system that reflects the full cost to society of everyone continuing to drive his own car to work.

Obviously, the greatest savings can be achieved by bicycling, or even walking for short trips, thus substituting human energy for petroleum. The bicycle boom is sweeping America, with some 30 million new bikes purchased in 1972 and 1973. Many owners would ride their bikes to work if they didn't have to risk their necks dodging automobiles. America could make no better investment in energy conservation than to build separate bicycle paths throughout its major metropolitan centers.

There is no cure-all to the problem of sprawl, but certainly planning the growth of urban areas to enable people to live in pleasant surroundings fairly close to their work must be part of the solution. "Come back to town" might well be a constructive citizen response to the energy crisis that would keep our cities from dying a slow death as well. A program is needed to revitalize our cities for people of all income levels and to build new work-oriented towns that include new residential communities in the decades ahead. With good planning, new homes and offices can be brought closer together, and the people living in these communities will not only save energy but have time to enjoy life as well.

To be sure, most people will be living and working in existing buildings for decades. That is why doubling the average mileage of automobiles and building improved public transportation systems are the essential elements of a solution. It is not necessary for people to give up the family cars. What is necessary is cars that go twice as far on a gallon of gas and public transportation alternatives be available to permit people to drive less and enjoy it more.

The automobile dominates passenger traffic between cities, accounting for 87 percent of passenger miles. Airplanes, which are a rapidly growing competitor, account for about 10 percent, with buses and railroads losing out year by year. As a result the energy required per intercity passenger mile is skyrocketing. It takes 10 gallons of gasoline per person to fly from Boston to New York City on a regular jet flight as compared to 7 gallons by car, assuming both vehicles are half full. If a passenger took a fast train or a bus, however, it would require only 2 gallons.

Ironically, rail transportation has become steadily more efficient in recent decades through dieselization, but it is being used less. At the same time, the automobiles and airplanes that have taken over intercity traffic have become more wasteful. Massive public funds have been used to subsidize and accelerate this wasteful growth by building new highways and airports, but we have spent very little money modernizing the railroads.

Airplane traffic will no doubt continue to dominate long-range transportation. But air traffic at major cities is already overwhelming existing airports and sites to build new ones are scarce. It is time to provide an alternative to the short plane trips that use energy so wastefully. Fast train service will become more competitive, due to higher oil prices and upward adjustments in the federally-sanctioned airline rates which now subsidize short flights of 200 miles or less at the expense of long-haul passengers. A combination of enlarged fast train service and full cost for short-haul flights would divert a significant amount of traffic from the air to the rails. It has already occured in the Washington-New York corridor, where a fast train is running.

Still, most people will continue to want the convenience and flexibility of their personal car for family and business trips between cities. As noted above, the main road to saving energy thus lies in cars that travel more miles per gallon of gasoline.

The electric car is often mentioned as an alternative, especially a small vehicle to be used to drive to the mass-transit station or for trips within a metropolitan area. An electric motor is more efficient than the internal combustion engine and an electric car would cut in half the energy consumption per passenger mile even if all the losses in generating electricity are included. The electric car would shift air pollution from city streets and highways to remote power plant sites. If the plant were nuclear-powered, it would mean little or no air pollution. The problem with the electric car is that the technology has not been perfected. Research talent has not yet been mobilized to create a lighter and more powerful battery that would give the electric car sufficient driving range to become an attractive alternative.

An electric car may never be feasible for trips between cities but an electric car that would have an economical and feasible range of 50 to 100 miles for metropolitan use is a real possibility. An investment of $10 to $20 million a year for a decade in battery research could well provide us with an alternative form of personal transportation that would not only cut energy consumption in half but also shift it from scarce petroleum to the fuels which will be more abundant in the future—nuclear-, solar-, or fusion-powered electricity.

A policy of energy conservation in transportation can look to communications technology for assistance. Widespread availability of closed circuit television would make some business travel unnecessary. If access to the company "tube" rather than the airline credit card were a badge of prestige, it could quickly catch on. People who are already acquainted might find that the larger-than-life screen provided them with a satisfactory alternative. Indeed the wired city of the future might eliminate much travel for banking, mail and some shopping. People enjoy each other's company, and no one suggests that future Americans would want to live a synthetic, push-button life. But much of the scurrying about in contemporary life could be eliminated. A more efficient and more relaxed environment will require much less energy—human as well as commercial.

Passenger travel is the main drain on our energy resources

—only about 25 percent of transportation energy is used to move freight. Trucks haul most of the traffic and largely in the metropolitan areas. Here again urban sprawl is a root cause of the increasing truck mileage for deliveries to homes, offices and industry. New communities would not only be more compact but underground pipeline conveyor belts and other more efficient systems such as pneumatic pipelines also could be installed to deliver freight.

Railroads still move twice as much freight between cities as trucks but the relationship in energy consumption is the opposite. It takes about four times as much petroleum to move a ton-mile of freight in a truck as in a rail car. Water transportation and pipelines are even more efficient than railroads. Moving freight by air is an energy extravagance.

The railroads have been steadily losing freight traffic to the trucks. A concern for energy conservation as well as pollution control and transportation economics would reverse this trend, certainly for all shipments over 200 miles. A program of public investment to modernize the nation's rail system coupled with elimination of the current subsidies to the trucking industry thus becomes an important part of a new energy-saving policy.

As this nation has moved from the railroad, to the automobile and truck and now to the airplane, we have accelerated both the pace of life and the drain on energy resources. And as automobiles have become larger and cities sprawled into the countryside, the demand for petroleum has continued to rise. Energy conservation in this area of our national life implies smaller cars, faster trains, more compact new communities, and more modern mass transit systems in existing metropolitan areas. The most promising route to energy savings lies in making our transportation system more efficient.

INDUSTRIAL ENERGY SAVING

More than 40 percent of total energy is consumed by American industry. The intensity of industrial energy consumption varies from a research laboratory, which uses energy much like a home or office, to a fertilizer plant, where natural gas is the raw material and accounts for half the cost of the manufacturing process. For industry generally, energy accounts for only about 4 percent of its value added.

American industry has emphasized using mechanical energy to automate its operations and save labor. In the factory and on the farm Americans have learned to mass produce food and goods with fewer people working shorter hours at jobs that require less manual labor. This is the result of using machines lavishly powered by energy. With bulk energy inexpensive there has been little incentive to seek ways to economize on the energy required to run the machines. Most industries have taken for granted the abundant availability of low cost energy and are sorely troubled by recent shortages and skyrocketing prices. These shortages have already shocked some industries into energy conservation programs and into new awareness of the economic facts of life: There is a trade-off between the capital investment needed to conserve energy and the savings in fuel bills that would result. As energy prices increase, mechanical devices that are available to use energy more efficiently will be bought and used. And as prices continue to increase, industry will invent new devices and adopt more fundamental energy conservation plans.

There is major scope for savings. Each industrial plant is enclosed in a building that requires energy for space heating, cooling, lighting and other functions in much the same way as an office building or a home. In the era of cheap energy, industrial managers were no more frugal than householders. Opportunities are thus available for reducing heating energy 40 percent by improving insulation and by operating and maintaining industrial plants more efficiently. One attractive new idea is the use of infra-red heating in high bay buildings, a technique which directs heat on the floor where it is needed and produces large savings.

It is difficult to be precise in discussing energy conservation opportunities in industry because no two industrial processes are identical. Each industry will require detailed analysis by experts familiar with the industry's techniques and technologies and their adaptation for more efficient use of energy.[179] It is possible, however, to identify the main functions that energy serves in industry and the major areas for conservation. The bulk of industrial energy consumption is used as process steam, direct heat or electricity to drive motors. These three categories account for about 90 percent of industrial energy use and more than one-third of total U.S. energy consumption.

Almost half of the energy consumed by industry is fuel converted into steam that is used as part of the manufacturing process. Process steam is 17 percent of total energy, more energy than is used in all the office buildings in the nation. The steam is produced in a very primitive way, simply by burning fuel, mainly because fuel has been cheap. With higher fuel prices, industry will see that the steam could be produced with far less consumption of fuel, perhaps half as much, through new methods based on the principle of the heat-pump, which uses the heat available in the surrounding air. And, as we will see in the next chapter, a major opportunity for energy saving lies in using the waste heat from central station power plants as a source of industrial steam. Such savings cannot be achieved at once, but a beginning can be made in the design and the siting of large, new plants. If such power plants could be located in or near urban areas, the waste heat could be used in existing industries. And indeed in building new communities in the future "district heating" systems could be installed so that the waste heat is piped to each home and individual furnaces are eliminated. These are big, long-term undertakings. A small and immediate one is to make certain that all steam pipes are insulated. In between these extremes there are many major areas for saving energy in the use of process steam.

There are numerous other measures that can be taken within existing industrial plants to use energy more efficiently. In many cases electricity is used as a source of direct heat which is inherently less efficient than obtaining the heat by burning the fuel on site. Reviewing production schedules and operating practices in industry could probably save more energy than equipment changes. Mechanical devices such as heat wheels and heat exchanges that reuse waste heat and prevent its escape through open stacks can produce big gains in efficiency.

It is not possible to quantify the net effect of the measures industry is already beginning to take to conserve energy. To a large extent it will depend on the fear of shortages and how rapidly the price of industrial energy rises. Nevertheless, if management is alert to the early payout on investments of time and equipment to conserve energy, it would seem quite feasible that overall savings of 25 percent can be achieved in existing plants over the next ten years.

For an industrial manager, shortages of energy generally present an economic problem very similar to a homeowner's —a sharp increase in a cost item that is less than 10 percent of his budget. But a handful of industries are very energy intensive and account for about 80 percent of industrial energy consumption. To them conservation may become a question of economic survival. One such industry is primary metals, the largest industrial consumer of energy, especially steel and aluminum, and to a lesser extent copper and zinc. The manufacture of hundreds of chemicals such as chlorine, soda ash and ammonia used for such diverse items as fertilizer and pharmaceuticals is the second largest area of industrial consumption. Another major consumer is the paper industry where large quantities of heat are used to convert trees into products such as newsprint and wrapping paper. The manufacture of cement requires large quantities of energy. The petroleum industry itself is a major consumer of energy in refining crude oil to gasoline and other products. Large quantities of fuel are also needed to process foods into the variety of pre-packaged items on the supermarket shelves. These industries have an acute sensitivity to the price of energy and are likely to feel the most pain as costs soar.

Very little public information is available concerning new conservation technologies in these energy-intensive industries. But the aluminum industry claims it can cut energy requirements by 30 percent in new plants. And comparable savings are technically feasible in cement, steel, paper and perhaps other industries. As energy costs rise, substitute products that are less energy intensive become economically attractive. In the past there has been little incentive to seek them out. Now, soaring energy prices should send industrial entrepreneurs searching for alternatives in areas where rising energy costs already account for 10 to 50 percent of the product's total cost.

There are many avenues toward savings in the energy-intensive industries. One is the increased recycling of metal, which conserves both energy and material resources. It requires only a tiny fraction as much energy to recycle and remelt metals such as iron or aluminum as is required to reduce new ore mined from the earth.[180] And far less energy is needed to clean and reuse glass bottles than to continue to make new ones. Recycling can greatly reduce energy requirements in large sectors of industry. For example, steel and aluminum alone account for more than 7 percent of total energy consumption.

The recycling of materials is by no means easy. A beginning has been made in copper where most metal is reclaimed and where 19 percent of production is from recycled metal. But there are obstacles. First, there is the practical problem that many of the metals to be recycled are now alloyed or chemically reacted with other materials in their manufacture. Physical separation is not only expensive but also would itself require considerable energy. Market incentives also work against recycling. New materials do not include society's cost of disposing of the waste as compared to reusing the material. The tax laws provide a depletion allowance for extracting additional minerals from the earth, but there is no comparable tax incentive if the same minerals are mined from the trash can. Therefore, low freight rates are available for metals and paper products but not for material to be recycled.

New government policies are needed to remove these artificial barriers to recycling. This issue is but part of an even larger opportunity for energy conservation in the industrial sector which we will discuss in the last chapter. There must be a trend away from building things to be used and quickly thrown away. Quality control and durability must be the prime objective of the manufacturing process.

SUMMARY

For many decades the U.S. has led the world toward a pattern of energy consumption in which increasing per capita use has roughly paralleled the increasing material affluence and mobility of society. The time has come to separate the two curves, so to speak. The efficiency of energy utilization can be improved, and our society can continue to spread the basic material comforts among the population without such rapid and wasteful growth in the consumption of energy.

In this chapter we have seen some of the many ways in which our society can "have its cake and eat it too." We have concentrated on the technical possibilities available for conserving energy and have seen that they are manifold. That they have not been adopted in the past has been due mainly to the fact that energy prices were so low that it was not worthwhile to incur the expenditures—sometimes modest, sometimes substantial—necessary to secure energy savings. Energy prices are now rising fast. This development sharply increases the benefits of energy conservation in relation to the costs. But it would be an optimist indeed who would assume that better economics can do it all alone. Some markets, particularly the markets for buildings and transportation, are full of imperfections which need to be overcome by direct action on the part of government.

Such actions are essential if we are to weave energy conservation into the fabric of our society (see Chapter 13). The rate of growth in the consumption of energy in the home, in the office, in automobiles, and in manufacturing plants can be reduced by one-third by adopting sensible measures that on balance might also improve the quality of life.

The consumption of energy in the U.S. in 1973 was the equivalent of 13 billion barrels of oil. Most current projections are that by 1985 requirements for energy will total the equivalent of 20 billion barrels. If we could implement the efficiencies documented here, however, a 2.0-percent-a-year growth rate, or even less, would be altogether feasible. United States' energy demand by 1985 would then be only the equivalent of 16 billion barrels, a 20 percent reduction. And if we look at the year 2000, the savings approach the equivalent of 13 billion barrels per year, as compared to business as usual estimates, an amount almost as large as our total consumption of energy in 1973.

It is important to realize that because of the existing stock of houses, machines and indeed our attitudes about energy, the full potential for conservation will not be achieved immediately. It is also fair to note that large savings will flow automatically from higher prices and increased consumer awareness. But the demand for energy still reflects many outdated government policies geared to promotion rather than conservation. Chapter 13 recommends a series of measures needed to supplement market forces and reverse this policy emphasis.

Adoption of these conservation measures and others in the production and conversion of energy could avert the coming energy crisis for several decades. The needs of society could be met during this century even though we would still face serious problems of pollution and proper allocation of scarce energy resources. Conservation measures would buy time—shortages now projected for 1985 would be deferred until the year 2000. If this added time were used to implement a massive program of research and development for cleaner energy resources, our high energy civilization could then continue without the specter of crisis. Without such conservation now, however, there probably is not time to avert a real crisis.

CHAPTER 11

Making Better Use of Existing Sources

There is an obvious and urgent need to clean up and enlarge the supply of energy. The basic question is whether science and technology offer real promise of meeting this need. What are the possibilities? And why haven't they been developed?

The short answer is that a vast array of options exist for improving existing sources of energy and perfecting entirely new sources. But as a nation we have sadly neglected their development. Neither the government nor the energy industry anticipated the combination of shortages, pollution problems and foreign policy concerns that now beset us.

In the past quarter century research and development in the energy field has been confined almost exclusively to atomic power. Through massive federal funding, we have succeeded in taming the atom for use in generating electricity. Even so, the truly efficient nuclear power plants—the breeder reactors—are still decades away from commercial reality. There is no expectation that nuclear power can supply more than a relatively small share of our energy needs in this century. And nuclear power itself has raised as many questions as it has resolved.

Past efforts to improve the mining and burning of coal have been virtually non-existent. The same pattern of benign neglect has produced little progress in learning how to use the vast shale oil resources, the heat in the earth, the sun, our solid and animal wastes, or other potential sources of energy. Without federal funding, the cold hard fact is that private enterprise has not done the job.

There are several lessons to be learned from our experience. Perhaps the most important is that the incentives are lacking for private, profit-seeking companies to invest the millions of dollars in long-term, high risk research that might enlarge the energy supply. We are not searching primarily for new patentable inventions that might pay off for the inventor and his backers. Rather, the process demands tedious and expensive engineering to test known concepts and build the hardware to make them work reliably and economically on a commercial scale. Investments in energy R & D take a decade or more before any pay-out, and even then more of the benefits go to the consumer and society than to the companies making the machines. We spent the 1960's discovering that private enterprise would not tackle this task. Now, hopefully we realize that federal funding is necessary.

Another lesson is that the development of new energy sources takes a long time—measured in decades, not years. And it takes an even longer time for new sources to begin to supply the major share of the market. The most optimistic projections for atomic energy, for example, still anticipate that fossil fuels will supply 60 to 75 percent of our total energy needs in the year 2000. Of course, this timetable could change. But it would require the sort of intensive effort the nation tends to make only in wartime or in achieving a glamorous objective such as sending astronauts to the moon.

A third lesson is that we have been unable to anticipate very well either the benefits or the problems of new technology. It was predicted that nuclear power would save consumers big money over conventional sources and solve the supply problem. Thus far the main reason we are building atomic plants is to avoid air pollution from fossil fuel plants, a reason which was all but overlooked in the original justification for nuclear power. And on the problem side of the ledger, there was little anticipation of thermal pollution, plant siting or other environmental concerns arising from nuclear power.

We must spend more time trying to anticipate and remedy the side effects of new technologies before we perfect them and begin relying on their use. Otherwise, in trying to solve one set of obvious problems, we run the risk of creating new ones that could be even worse. The potential problem areas are by no means confined to environmental concerns. For example, the development of new energy sources should be sensitive to the problems of monopoly and the need for public control. We must be alert for technical options that might bring new competitors on the energy scene as contrasted to intensifying the need for government control of existing companies. Research and development options should be chosen with an eye to preserving the political power of the people as well as providing them with the power to run their machines. And we cannot lose sight of the real costs to society. If energy becomes increasingly expensive it will be less a useful servant to mankind.

Albert Einstein once said: "The concern for man and his destiny must always be the chief interest of all technical effort." The technology to solve the energy crisis should be designed with the broadest possible vision of its ultimate impact on the lives of people.

Energy supply is in a state of transition. There is already growing recognition that the fossil fuel age will reach its peak in a few decades and that atomic energy, hopefully soon joined by less frightening partners, will supply the bulk of this planet's energy needs a half century from now.

But the fate of the world to some small (or perhaps quite large) degree depends on how we handle the transition. A two-pronged technological effort is needed. We should make fuller and more flexible use of remaining fossil fuels and eliminate much of their adverse impact on the environment. In a parallel and overriding effort, we should perfect the altogether new and self-renewing sources that hold promise of being relatively "clean," recognizing all the while the elusive nature of "clean energy."

In this chapter we shall explore how to make better use of

what we have and then survey the exciting opportunities for eternal sources of energy.

FUEL LEFT IN THE GROUND

Industry digs and drills for energy to earn profits under the economic rules that prevail. An oil company leaves two-thirds of the oil in a reservoir, but it does so only because it's cheaper to move on to the next reservoir. In the past, prices have not permitted the profitable recovery of the oil that remains. Industry therefore "skims the cream" rather than spending the extra money required for secondary and tertiary recovery of oil. The sad fact is that if the remaining oil is not recovered while the field is still producing as a practical matter, it is lost forever. For once the flush production is completed and a well is capped and abandoned, the physical problems and enormous costs of reopening it mean that the remaining oil has been written off.

Technology is continually being developed to recover more of the resource and leave less in the ground. The industry could produce twice as much oil as presently contemplated from the oil fields already discovered. Oil men could also learn to recover some of the heavier oils that total some 300 billion barrels of potential resource. Perhaps a recovery rate of no more than 10 percent is possible with these heavy oils, which are found on land in California and in the Southeast. But even a 10 percent rate of recovery would mean some 30 billion barrels, which would nearly double the nation's existing proven reserves.

The National Petroleum Council and the Interior Department estimate that the U.S. will be consuming some 25 million barrels of oil every day by 1985 (without the conservation measures discussed in the preceding chapter) and only half of that amount will be produced in the U.S. Certainly an extra 2 million barrels of oil per day, and more, could be produced by 1985 from what is now being left in the bottom of the barrel and from the heavy oil that is being overlooked.[181] Such a program could make as large a contribution as the Alaskan oil toward alleviating problems of energy supply until new sources can be developed. But prevailing prices and tax laws inhibit the development and use of the technology to produce this admittedly more expensive oil "at the bottom of the barrel," which nevertheless has value to society for environmental and foreign policy reasons. Unfortunately, these values are not reflected in federal programs of price controls, tax laws or leasing policies. There is no program aimed at tapping these readily available and environmentally superior sources of energy.

In contrast, recoveries from producing reservoirs of natural gas are high—90 to 95 percent is recovered. Even so, the additional 5 or 10 percent could mean the difference between shortage and an adequate supply in the years ahead. There is no physical or technical reason why extraction could not be close to 100 percent if the right combination of industry incentives and governmental requirements prevailed.

The major waste of natural gas, however, has occurred mainly above ground where it is sometimes burned as a huge flare right at the well. Such flaring occurs where natural gas and oil are in the same reservoir and are produced together before there is a pipeline connection and a market for the gas. Waste dominated the early history of gas production in the United States, and it still prevails on a massive scale in such nations as Saudi Arabia and Venezuela, which have no market for their natural gas. Even in the United States, where there is an acute shortage of natural gas, the federal government still permits flaring on the Outer Continental Shelf. Enough gas to meet all the needs of Washington, D.C., was flared in this manner in 1971. The government can eliminate this waste by requiring that the natural gas be reinjected into the reservoir until it can be sold.

Coal is also wasted in the ground in huge quantities. The prevailing practice is to recover only half the coal in an underground seam. The other half is left as supporting columns and because it is too expensive to mine out. This practice is perhaps understandable in view of the abundant coal resources as contrasted with petroleum where shortages already exist.[182] But it was not many decades ago that petroleum also seemed abundant. It is by no means too early to begin to practice conservation of our coal resources. Large blocks of coal strategically located with respect to markets in the eastern U.S. are already scarce.

Coal recovery could be dramatically improved with new mining methods. The longwall method now used in Western Europe recovers about 85 percent of the coal in place and is inherently much safer for the miners as contrasted to the "continuous mining" technique used in American mines. The longwall method provides roof support for the miners and then automatically collapses the roof to obtain more complete recovery. By contrast, continuous mining, which is not continuous at all, merely claws out the easiest obtainable coal and leaves the remainder as supports. The longwall method offers a way to improve both the efficiency and safety of coal mine production. The Federal Bureau of Mines could require all new mines to use the longwall technology, which would be justified on safety grounds alone.

NATURAL GAS IN TIGHT FORMATIONS

There are large volumes of natural gas in the ground in known formations—some 600 trillion cubic feet—that are too "tight" to be economically recovered using conventional means. The rock formations in which the natural gas is located are too dense to permit much of it to flow naturally to the surface. One possible method for recovering this gas is to use low-yield nuclear explosions to fracture the rock. A nuclear charge is lowered through a well hole deep into the ground, and then the well is sealed and the charge detonated. The resulting underground blast does not release radioactivity into the atmosphere but it pulverizes the surrounding rock, opening up a big hole and creating thousands of tiny cavities in all directions into which the natural gas in the reservoir can flow and be recovered.

The nuclear stimulation method has already been tested successfully in New Mexico and Colorado. In 1970, Project Gas Buggy detonated a 20-kiloton device—about the size of the Hiroshima bomb—in a dense gas-bearing rock formation without releasing radiation to the atmosphere. In the first two years of operation, wells drilled around the blast site have yielded about 300 million cubic feet of gas—a tiny volume compared with the annual U.S. consumption of 23 trillion cubic feet, but a sizable amount considering that without the Gas Buggy explosion the production potential of the field was next to zero. In May, 1973, a similar experiment took place in Colorado. This, too, was successful. No radiation leaked into the air and a sizable flow of natural gas was freed for recovery. The economics of nuclear stimulation of tight natural gas formations is still uncertain but it appears rather attractive on paper, especially as compared to imported natural gas on the West Coast costing some $1.50 per thousand cubic feet. Estimates of the prices of fuels from stimulated wells range up to 75 cents per thousand cubic feet.[183]

The problem is not technical feasibility or economics but rather public concern over the radioactivity produced by underground detonations. Underground nuclear blasts in the number required for sizable production—up to 240 per year[184]—face strong opposition from the people in the Colorado-Wyoming-Utah region where they would occur. There is fear of radiation leaking into the air or seeping into the ground water. Although the blasts would be made with relatively "clean" bombs they would in fact result in a tiny amount of radiation in the natural gas. A person cooking with such gas would encounter very low radiation levels, well below permissible standards, but a measurable quantity nonetheless. And radiation is cumulative in effect which could pose serious problems. There is also concern that a series of blasts might trigger an earthquake. The shots are in fact felt in the surrounding area and sometimes cause minor damage to homes.

For these reasons nuclear stimulation is moving ahead very slowly. Another option that should be explored is the fracturing of rock by hydraulic means. Hydraulic fracturing has been used for many years with little or no adverse effects. It involves drilling a conventional well hole and then pumping in fluid under high pressure until the surrounding rock fractures. Many more wells would be required in a hydraulically stimulated field, but it would avoid the fears and very real dangers of exploding hundreds of atomic bombs. And hydraulic fracturing now also appears to be economical.

Stimulation of the tight gas formations could start bringing more natural gas into production in a few years. It would take a decade to achieve sizable productive capacity, yet recent studies have shown the potential could build up to 2 to 5 trillion cubic feet per year by 1990.[185] This source of natural gas could be an important supplement during the next two decades when conventional supplies will be very scarce.

CLEANER ENERGY FROM COAL

Along with nuclear power, coal holds the key to meeting our energy needs for the rest of the century. These two power sources share the same dilemma: Although they now represent our most abundant energy forms, they pose serious environmental problems. In fact, in the minds of some (but perhaps different) segments of the public, both coal and nuclear power are so environmentally disruptive that they should be eliminated altogether as energy sources. But with respect to coal we have yet to make a truly determined "clean-up" effort.

We have discussed the problems of coal mining and potential solution. Deep mining that is relatively safe need not be only a dream. Sound reclamation practices, rigorously enforced, could permit strip-mining to continue in some areas. Actually the major bottleneck at the moment is the lack of technologies enabling us to burn coal cleanly.

Giant scrubbers and other techniques can be perfected to remove the sulphur and perhaps other pollutants from coal as it is being burned. Some research and development has been funded in this area, and it is on the verge of achieving important results. The technology to "scrub" pollutants out of discharges from the coal-burning process will be expensive. But such technology could make it possible for coal to supply fuel for electric power plants and large industries in compliance with air pollution standards. With a strong push, "clean coal" is only a year or so away from commercial feasibility. In fact the technology may already be "available," if the utilities knew for certain they would be required to use it.

The fluidized-bed boiler is an altogether different approach to burning coal more cleanly. It is a new type of boiler, simpler in many ways than current boilers, with a layer of limestone or other particles making up the bed at the bottom. Pulverized coal is burned in the boiler using injected hot air, and the sulphur and other pollutants are absorbed by the bed of limestone pebbles. A main advantage of the fluidized-bed scheme is its thermal efficiency: more than 50 percent compared with conventional coal-fired plants, which are about 40 percent efficient. Also, because the new boiler is a fairly straight-forward design compared with existing boilers, it may actually save money. Electrical energy made from coal through this process may be cheaper than other options for cleaner energy.[186] Several years ago, a small pilot plant using a fluidized-bed boiler was built with government research funding and successfully operated. The next step is to demonstrate a larger scale plant; if it is successful, then the process can be used commercially. But the program has been starved for money and the larger demonstration plant has not yet been completed.

It is important that we perfect a variety of ways to burn coal in its natural form. Coal—as coal—is cheaper than its conversion to a gas or a petroleum liquid. And the technologies to burn it as mined can be perfected soon, by 1980 or earlier with a determined effort.

The prospect of being able to use our enormous coal resources to fill the petroleum gap is of course very appealing. It may be the answer to America's energy problems. But it also may be as big a disappointment in the 1980's as atomic energy is to some people today.

The federal government has underway a program of pilot-plant tests of various technologies for converting coal into a synthetic gas or oil. The end products are the same as natural gas or oil except they will be considerably more expensive. They also raise environmental problems of their own. Participants in the program are optimistic, as they should be. But thus far the pilot plants have encountered troubles, and there is an uneasy feeling that the whole effort lacks a strong technological base. We may well face the same situation as with atomic power in the 1950's, when many experiments had to be closed down as failures.

Coal research has been so poorly funded in the past that not enough is known about the basic chemistry of coal, much less about engineering the complicated hardware needed for large-scale commercial operation. Project Gasoline—dedicated in 1968 and yet to produce a gallon of gasoline from coal—stands in Cresup, West Virginia, as proof that failures are a very real possibility in this risky field. Of course there is no doubt that in a real pinch coal can be converted to petroleum. The Germans did it in World War II. But can new processes be perfected to do it much more economically and at the same time remove the sulphur and other pollutants? Sooner or later success is certain. But because of our past neglect it may be later. And it could be never if environmental and other problems cannot be resolved.

Western coal is the most likely source of a new coal-to-petroleum industry. A first hurdle is strip mining it. The jury is still out on the question whether these arid lands could recover from massive stripping. The key bottleneck might turn out to be water. A large synthetic coal industry in the West would require huge quantities of cooling water for the plants and municipal water supply for the people who would move into the area to work. There is insufficient water in the region to support such growth and no effort underway to supply it. In the end the basic question may be whether inhabitants are willing to accept the drastic changes in their region that such large-scale development projects entail.

OIL SHALE—THE SLUMBERING GIANT

Suppose we discovered a giant oil field in the heartland of America—one with hundreds of billions of barrels of oil to satisfy our needs for decades. Would our energy supply problem be solved? Not quite. Amazingly, such an oil field exists. It was discovered by the Ute Indians more than a century ago in Western Colorado and in adjacent areas of Utah and Wyoming. Long before the first settlers came to that Rocky Mountain region, the Utes were using this oil for their campfires.

This sleeping giant is our shale oil resource. Shale oil is not a free-flowing liquid as in a natural oil reservoir. The oil, known as kerogen, is trapped in solid form as an integral part of the shale rock. The recovery of oil from oil shale is actually a relatively simple procedure. The conventional approach is to mine the rock and heat it in a vessel called a retort to temperatures of 800-1,000° F., thus vaporizing the oil, which is then condensed and drained off. There are three variations on this basic process, and the most promising appears to be the "gas combustion retort," which has the advantage of requiring no cooling water.[187]

The potential volumes of oil from oil shale are tremendous. According to the Interior Department's Bureau of Mines, the amount of oil that can be obtained from the higher grade shale ores (25 gallons per ton or more) is estimated at around 600 billion barrels. This is roughly ten times the potential of the rich North Slope of Alaska. The Bureau of Mines says that there are also another 1,200 billion barrels available in lower grade ores.[188]

The shale resource is located in a fairly concentrated area, and almost 80 percent of the land is owned by the people of the United States. At present there is no commercial production of shale oil in the U.S. even though shale oil industries have developed in other nations, including Russia and China. Shale oil has not yet been developed in the U.S. because oil supplies have been available in conventional form at lower costs to the industry. With domestic oil supplies getting tighter, however, interest in the resource has begun to stir.

The Interior Department leased six tracts in Wyoming, Colorado and Utah in 1973. These tracts alone are sufficient, the Interior Department says, to allow a production level of 1 million barrels per day by 1985.[189] The government estimates that the shale oil industry could reach a production level of 2 million barrels per day by 1985, which would be 8 percent of estimated domestic demand in that year. Production then could expand rapidly if bottlenecks were removed.

There is no assurance, however, that shale oil production will move ahead on any timetable. Leasing the land is a far cry from ensuring the development of shale oil. Even if industry does move into commercial production, it probably will be at a slow and cautious pace. There are serious environmental difficulties as well as economic uncertainties.

Shale oil production is undoubtedly profitable at current crude oil prices, but there is still uncertainty as to the level of prices five to ten years hence when the first big plants could be completed. Future prices are intimately connected with government policies, and there is the added financial risk inherent in bringing any new capital-intensive industry into existence. Such risks are compounded by uncertainty as to how costly the environmental problems will turn out to be, or whether they can be solved at all.

There should be no illusions that we can rely on a highly competitive energy industry aggressively to develop this new resource. Thus far there has been no rush into shale, although plans have been announced to build the first commercial-sized plant at a cost of $250 million which would produce 46,000 barrels per day. The plant would be jointly owned by Atlantic Richfield Oil Company and the Oil Shale Corp. A major constraint up to now has been the federal government itself, which owns most of the good shale lands. There is still much uncertainty over the future availability and terms of federal leases. At present each company is limited to leasing 5,120 acres, which inhibits a large gamble by a single company because it cannot acquire enough land for a major development program.

Industry has also shied away from shale oil because the tax laws have been stacked against shale. The federal tax laws have provided a large break to conventional oil through the 22 percent depletion allowance and expensing of intangible drilling expenses. For shale oil, present law allows only a 15 percent depletion allowance. Due to the capital-intensive nature of shale oil exploitation, intangible expensing is of lesser importance. Giving shale oil a 10-15 percent handicap as compared to conventional oil has discouraged its development.

Moreover, shale production is in a sense a new business for the oil companies. Their expertise in finding petroleum is not useful. The shale is there. It is a mining and manufacturing operation, which once it became a going industry could easily be entered by newcomers. The oil companies would have no special advantage.

Perhaps the key question that controls the green light for shale oil is its impact on the surrounding landscape. The component issues are the land-use problems of mining and disposing of the shale rock, contamination of the air and water from mining and retorting operations, and the basic conflict between industrialization and preservation of the natural environment. This array of problems must be weighed against the problems of alternative sources of petroleum. Shale oil development is land-based and thus would avoid the dangers of oil spills on beaches and oceans created by imports and off-shore drilling. And it would of course avoid the foreign policy and monetary concerns raised by importing oil from the Middle East.

The land-use question is obvious and formidable. Since every barrel of shale oil generates about 1.3 tons of waste material, the disposal problems are enormous. The spent shale cannot simply be returned to the ground, because it actually occupies a greater volume than the untouched shale ore did before processing. Even a 1-million-barrel-a-day shale oil industry would pile up about 475 million tons of waste per year. The rock could be used to fill in canyons and ravines but this could upset drainage patterns. Experiments are also underway to grow grasses on the deposits of processed shale.

One development that would minimize the impact on the land would be an underground method of mining shale. Some shale ore, of course, can be mined by conventional underground methods, but research is also being conducted on a more promising system called *in situ* (in place) processing. The *in situ* process amounts to igniting a fire in the shale formation that burns underground, thus separating the oil.

As the fire spreads, wells are sunk in sequence to collect the newly-separated oil. Holes are also drilled behind the area of combustion, through which hot gases are pumped to keep the reaction going. Scientists have also suggested detonating small atomic bombs underground to break up the shale formations and free the trapped oil. But this technique requires overcoming the dangers of contamination of ground water and limiting the radioactivity of the shale oil, which lies much closer to the surface than the tight-trapped natural gas reservoirs in the Rockies.

Research and development into ways of extracting the shale without reshaping the surface should be high on the list of priorities in any program for shale oil. The beginning of shale development, however, need not await perfection of the *in situ* method. The waste disposal problem will not become a major concern until the industry begins large-scale production. By building a commercial-scale demonstration plant as an environmental "yard stick," it is even possible that the methods for disposing of waste rock in large canyons and growing vegetation on the waste material could prove satisfactory.

velopment of shale oil. A 1-million-barrel-per-day shale oil industry would require between 100,000 and 150,000 acre/feet per year for processing, retorting, and refining the product.[190] This is equivalent to a pool of water approximately 1-1/2 miles square and 100 feet deep. Such a volume would claim a large share of available water supplies. Governmental action to remove the subsidies for water now going into irrigation would be necessary. If not, technology to desalt brackish water would be required to prevent a serious drain on the Colorado and Green Rivers, which flow through the shale oil areas.

There is an old expression in the West: "Water travels uphill to money." But if shale oil is to become a major source of domestic supply, new water policies and new projects will be needed to transport water to the producing areas from other watersheds. The birth of the shale oil industry will require a corresponding massive program of water resource management.

The final obstacle is the growing disenchantment with large-scale development of any kind in the Rocky Mountain area. The slogan "Don't Californicate Colorado" sums up the sentiment. If it prevails, shale oil production will be impossible. In any event, serious attention must be given to the secondary effects of shale oil development in a region that has escaped the tumult of industrialization.

To awaken the shale oil giant will require clearing major hurdles. Yet the nation's huge reserves of shale oil would support a massive production effort for decades once the industry gets underway. Perhaps the biggest obstacle to the development of the industry is not economics in the narrow sense but uncertainty over government policies and the attitude of the public on environmental issues. Government action will be necessary to assure speedy construction of the first commercial-sized shale plant in a manner that will demonstrate whether the environmental concerns can be satisfied. We need to know—and quickly—whether we can count on the huge potential of shale oil to ease our energy supply worries.

URBAN WASTE—LOOKING IN OUR OWN BACK YARD

One of the most frustrating problems of our industrial society is the disposal of the enormous quantities of solid waste that we are continually piling up. Currently in the U.S. each year, more than 250 million tons of cans, bottles, bones, old tires and other discards are burned, buried, or otherwise dumped into the environment in ways inconsistent with conservation or ecological goals. This figure includes only what is commonly known as "trash." If we include the enormous amount of waste products from agriculture, industry, and that other household waste material—sewage—the total reaches an astounding 2 billion tons of waste each year.[191]

In the face of energy shortages it is amazing that, although we know how to convert much of this material to synthetic oil and natural gas by a fairly simple process, we are not yet doing so on a major scale.

The energy potential of solid waste is derived from its organic components, mainly cellulose (paper and wood) and food wastes, as well as lawn and garden debris. These materials, scientists have discovered, will yield a combustible oil or gas if heated in an airless chamber. The process, called "pyrolysis," in addition to contributing to the nation's energy inventory, may prove the best solution to our growing solid waste disposal problem.

The possibility of converting solid waste to synthetic fossil fuels was discovered several years ago through research by the government on producing synthetic oil from coal. Researchers at the Interior Department's Bureau of Mines found that if coal is heated to 300-400° C. in the absence of oxygen, it will liquefy into a combustible oil. It was also found that any organic material subjected to the process will yield varying ratios of oil and combustible gas. Perfecting this process to turn solid wastes into energy would be a godsend to our populous metropolitan areas, which are currently burying refuse at considerable expense. By ceasing to "waste" our waste, we save the cost of disposing of great mountains of material, which yields no benefit to society other than keeping trash out of sight.

Let's examine the prospects, then, for increasing the nation's energy supplies by using solid waste as a supplementary fuel. Of all the material produced as waste in the U.S., urban trash seems to offer the best potential for serving as an energy source. Municipal refuse represents only about 10 percent of the total yearly U.S. waste product.[192] But it has the advantage of having been collected at centralized locations, where it would be relatively easy to build processing plants.

The Bureau of Mines has several pilot plants already in operation, and the results are encouraging. Plants are under construction or already in operation in St. Louis, Baltimore, and San Diego, as well as a few smaller cities. Urban refuse is collected normally and deposited at the processing plants. Either oil or fuel gas is produced, and the resulting fuel is either burned on the spot to produce steam for electric generation or sold directly to local customers.

This approach seems so eminently sensible, one wonders what the catch is. As usual it is economics. At present it is still quite expensive to convert trash to oil. The plant must separate tin cans, aluminum, and other trash from the organic wastes that can produce oil. The conversion process itself is also costly. The pilot plants in operation have been financed in part through grants by the federal government and are administered by the Environmental Protection Agency. Federal money has been combined with investments by the electric utilities, which will be the main customers for the fuels produced by the plants.

These plants will provide the first accurate information on conversion feasibility and cost. Even so, the demonstration plants are not small. The Baltimore plant will convert half the city's trash to fuel gas. A ferrous-metals recovery operation will accompany the energy conversion system and revenues from all products are expected to be around $1.5 million yearly.[193]

The Bureau of Mines estimates that the largest 100 metropolitan areas offer the best prospect for operating full waste-treatment/energy production facilities because the raw material would be readily available in sufficient quantities.[194] The organic material available from these areas now totals about 75 million tons per year, and the figure is expected to reach 100 million tons by 1980.[195] With present technology, this volume of organic feedstock would produce some 100-150 million barrels of synthetic oil a year, or about 2 percent of national needs. Even more promising is the potential for converting the waste material to fuel gas. The same amount of organic waste could produce up to 1.36 trillion cubic feet annually—or about 4 percent of current U.S. consumption. Moreover, the gas would be produced in the same metropolitan areas where it is in greatest demand, thus eliminating transportation costs.

Solid wastes also could be used as boiler fuel in power plants. In fact, the average grade of garbage has about one-half the btu content of coal and is a low-sulphur fuel.[196] Small electric power plants can use wastes as a fuel, and in the future such plants will no doubt be built. Solid wastes—

what we call trash—used as power plant fuel could supply more than 5 percent of U.S. energy requirements, the equivalent of more than 100 million tons of coal a year.

Solid wastes are a relatively small source of fuel as compared to agricultural and animal wastes, which have the potential to satisfy half our total energy requirements. The difficulty here is that the animal and agricultural wastes are widely dispersed and thus prohibitively expensive to collect. But the recent trend toward large animal feed lots, while it has created a pollution problem, has also concentrated a potential source of energy. Cow manure could be converted into crude oil or gas in the same manner as the organic wastes in urban trash. As with so many other options, we lack hard data concerning economic feasibility. The costs are doubtless higher than accustomed prices of petroleum but using concentrated sources such as the animal wastes in large feed lots could be economical if we make an appropriate allowance for the value of abating the pollution of our rivers.

The trend of agriculture is toward greater concentration and thus the potential for large-scale commercial energy from animal and agriculture wastes deserves a strong development effort. If this idea does prove feasible, it should be pushed into the commercial mainstream. The energy potential is 10 times as large as from urban wastes.

NUCLEAR POWER

Decades of scientific research culminated just over thirty years ago in the scientific breakthrough that began the atomic age: The construction of the first atomic reactor built secretly in Chicago during World War II—the famous Manhattan Project.

Research and development to use the atom for peaceful purposes began immediately after the war amid great expectations that atomic power would someday lift the standard of living of the entire population of the world. Scientists speculated that atomically-produced electricity would be cheap and abundant throughout the world. They foresaw automobiles that would run for months on tiny atomic fuel pellets and nuclear pumping stations that would convert seawater into fresh water and turn the deserts into gardens through irrigation.

We are now more sober in our assessment of atomic energy than we were a quarter-century ago. It is not quite the panacea for world problems that we expected it to be. Even so, the development of the peaceful atom to produce electricity has come a long way. The question is how much of the load can it safely carry in the future.

Fission Reactors

All nuclear power plants currently operating in the U.S. are powered by light water fission reactors. These reactors produce large amounts of heat. The heat is used to boil water that becomes steam, which in turn drives a conventional turbine generator to make electricity. The major difference between this system and a conventional generating plant is that the fuel being "burned" is uranium in a nuclear reactor instead of petroleum or coal in a boiler. And the byproducts are radioactive wastes, rather than gases and particles from fossil fuels which pollute the air. The reactors are called "light water" reactors because they use water as both the cooling medium and as a moderator in the reactor core to keep the fission process under control.

About half of the new electric power capacity under construction in the U.S. is nuclear. By 1980, nuclear power plants in operation are expected to have a combined generating capacity of about 100,000 megawatts. This figure will equal roughly 20 percent of U.S. electric power capability. In the 1980's nuclear power plants are expected to supply most of the new capacity.[197] Atomic power plants are being counted on as a vital part of the U.S. energy resource base.

The major problem with the atomic power plants under construction, aside from safety concerns, is that they waste so much energy. As we saw in Chapter 5, the world's uranium will support only a few decades of expansion with the current nuclear plant because only 2 percent of the energy potential is used. Furthermore, the light water plants must operate at lower temperatures and pressures than fossil-fueled plants, with the result that only 30 percent of the heat energy produced by the inefficient reactor is converted to electricity. The other 70 percent—an enormous quantity of heat—is wasted in the atmosphere.

We are entering the atomic age with a primitive technology, measured by any yardstick of efficiency. We are converting much too little of the energy potential into useful work for mankind.[198] There is a critical need for development of more efficient atomic power plants that will make better use of the uranium fuel and convert more of the heat into electricity.

A slight improvement over the current light water reactors in terms of efficiency is the High-Temperature Gas-Cooled Reactor (HTGR), now under development. The HTGR reactor uses helium under high pressure as the heat transfer medium, rather than using water to make steam. This type of nuclear plant uses the uranium somewhat more efficiently and would increase plant efficiency from 30 percent to about 40 percent.[199] It also can use thorium, a source of atomic energy perhaps more abundant than uranium. The plant can probably be designed to operate at even higher efficiencies when combined with a gas turbine. It could then operate without cooling water, which would make it feasible for use in arid areas such as the southwest.

The first commercial-scale HTGR, a unit of some 500-megawatt capacity, is near completion at Fort St. Vrain, Colorado. It has been delayed and plagued with operating problems. If it can be operated reliably and economically, the HTGR will add a more efficient new member of the atomic power team. Even so, the plant would not be a major advance, and efficiency in the use of the uranium or thorium would remain at about 3 percent.

The Nuclear Breeder

Another type of atomic fission reactor offers a truly abundant fuel supply. Called a "breeder" reactor, it does precisely what its name implies. It "breeds" or creates more usable atomic fuel than it consumes in the process of producing heat to make electricity.

The breeder reactor is not a perpetual motion machine. But it is little short of miraculous. It uses almost all of the 98 percent of the energy potential of uranium that existing reactors waste and converts more of the heat into electricity than existing atomic plants. And uranium is not the only potential fuel for the breeder. Thorium is also a potential "fertile" fuel in breeders, and its energy potential is even greater than uranium's.

If the breeder is perfected, the magnitude of energy available in the U.S. (and in other nations) is *several thousand times* as great as our petroleum resource and even hundreds of times as great as the vast coal resource. The breeder alone could keep a high-energy civilization going for centuries. The breeder not only would bring in an age of abundant atomic power, it also holds promise of lower-priced electricity than current technology provides. True, we no longer dream of ripping out the meters and having free power, as some enthusiasts did a quarter-century ago. But the breeder uses fuel so efficiently that it will no doubt reduce fuel costs to a very low level. The real question is how expensive the breeder will be to build. No doubt at first the plants will be very expensive, but as they become standardized there is reason to hope that the breeder will place a reasonable ceiling on energy prices and stabilize them at a level near current prices on a real cost basis.

The U.S. research and development effort on the breeder, in contrast to almost every other such undertaking, has been well-funded, and thorough. It has experienced some failures, as an R & D program must. The most prominent failure is the Fermi plant at Detroit, Michigan, which the electric power industry attempted to build before the "engineering homework" was completed. The breeder has been the primary objective of the AEC's atomic power research program from the beginning. The AEC has now learned enough from laboratory experiments, test facilities and detailed engineering design to proceed with building a large-scale demonstration plant that will produce some 400 megawatts of electricity. Construction on this first commercial-scale breeder plant in the U.S. will be begun in Tennessee in 1974 or 1975. The plant should be producing electricity in 1982 or soon thereafter. The Soviet Union, Great Britain and France, are building commercial plants that they claim are much nearer to completion. In fact the Soviet plant may already be operating on an experimental basis. Other governments have decided to go ahead with plant construction without waiting, as the U.S. has done, to complete elaborate testing of reactor components.

It is a race between the tortoise and the hare, so to speak. The U.S. expects to have the first breeder technology that will operate reliably and safely, but we may lose the race. More likely, it will be well into the 1980's before U.S. utility companies are satisfied that they should place an order for a commercial breeder. Even then the breeder will not gain instant dominance. A decade will be needed to build the first commercial plants. The year 2000 and beyond is the earliest period when the breeder might start carrying much of the U.S. energy load.

Even this timetable assumes swift progress in the demonstration plant program. But the breeder is not universally loved. On the contrary it is fast becoming the focal point of powerful and determined popular opposition to atomic power.

The breeder's success obviously would mean much wider-scale use of atomic power and thus would magnify the public concerns with current reactors discussed in Chapter 4—fear of a catastrophic accident, disposal of radioactive wastes, safeguarding nuclear material against blackmailers or sabotage, and thermal pollution. Actually the greater efficiency of the breeders cuts down on the thermal pollution of water by about one-third. The radioactive waste problem remains roughly the same whether current plants or breeders are used. And the safeguard problem could be greatly reduced if the breeders (which will be built in large-scale plants) were located in "energy parks" with a reprocessing plant on the same site to avoid the need for off-site transportation and with strong security measures in effect.

In addition to enlarging the scale of familiar concerns through expanded use of atomic power, however, the breeders pose two worrisome new problems.

The breeder highlights the public health problem of plutonium. The breeders will use as fuel the plutonium which light water plants now produce and store. For a comparable output of electricity, however, a breeder will cause about four times as much plutonium to be moved back and forth through the fuel cycle. Plutonium is a deadly poison and can cause cancer in very small quantities. The fear is that with so much plutonium being recycled the chances of release into the environment will climb sharply.

If the reprocessing plants could be located on the same site as the breeder plants and strict controls enforced, the plutonium could be contained. Nevertheless, many fear that the environment may be unable to tolerate even the smallest quantities of this man-made poison.

The other problem is the safety of the breeder plant. Fear of a plant catastrophe looms large in many scientists' minds even though the avoidance of accidents is the overriding design criterion of the breeder program. The breeder will operate at higher temperatures and pressures than current plants and will contain in its core large quantities of deadly plutonium. If the cooling system should fail and a meltdown of the core occur, the plutonium released would be equivalent in radioactive damage of the dropping of a large atomic bomb. The fact that liquid metal, a known fire hazard, is used as a coolant adds to the concern over plant safety.

These problems are understood by the breeder plant designers, and there has been more research work devoted to making the breeder immune to accidents than in any safety program in the history of technology. Yet in the final analysis the breeder's safety and the containment of plutonium will depend both on the foresight of the designers and on flawless quality control in the construction and operation of the plants and fuel cycle. The world has a right to be somewhat nervous about technology that depends on near-perfect engineering to avoid a nuclear catastrophe. Even if the odds against an accident are a million to one, the breeder still falls short of being the perfect solution to the needs of a high-energy civilization.

MORE EFFICIENT ELECTRIC POWER SYSTEMS

Each year a higher percentage of U.S. fuel production is converted into electricity. By the end of this century more than half of the fuel produced will be used to feed electric generating plants. Yet the generation of electricity wastes more than half of the fuel that goes into the plants. This is true despite decades of progress in which the kilowatt output per unit of fuel had been increased substantially.[200] But the limits of improvements attainable with current technology have been reached and in recent years power plant efficiencies have actually declined as less efficient nuclear plants have come on line. The energy currently being wasted by electric power plants equals the energy equivalent of about 3 million barrels of oil per day. That is more energy than we expect to receive from Alaska, and it is double the imports received from the Middle East in 1972.

We can cut down on this energy waste through improved technology for conventional generating systems and through new systems such as the fuel cell, which will use less fuel per unit of power output. And we can also learn to put some of the generating plants' waste heat to beneficial use.

One technical option that may be available in the near future is the combined gas turbine-steam turbine generating system. The concept is quite simple. A large conventional turbine generator is run on natural or synthetic gas, but instead of dumping large amounts of waste heat in the water, it uses the heat to boil water and drive a secondary steam turbine. The combined system achieves 50 to 55 percent efficiency in converting basic fuel into electricity, as compared to a maximum of 40 percent efficiency in more conventional plants.

The idea is not a new one. But when fuel was cheap and abundant, there was no incentive to use this more costly system in order to save fuel. Recent shortages and price increases have made this an economically attractive concept. The problem now is that many electric companies are finding it difficult if not impossible to buy natural gas for such new power plants. Synthetic gas made from coal is the answer. And if the gas is made at or near the power plant it can be a low btu gas, which uses a simpler technology than making pipeline-quality gas from coal. The remaining problem is to remove sulphur and other impurities so that the low btu gas is a clean fuel. Federally supported demonstration projects are underway. Although the effort has suffered from a slow start and inadequate funding, it could produce commercial hardware before 1980.

Another concept for more efficient power generation is magnetohydrodynamics, better known as MHD. The MHD process is basically as simple as moving a wire through a magnetic field to make it a generator of electricity. Instead of a wire, fossil fuels burned in compressed air form a hot stream of conducting gas that generates electricity as it passes through the magnetic field. This approach, combined with a conventional steam turbine, can convert 55 to 60 percent of the fossil fuel energy into electricity and thus offers an opportunity substantially to extend the life of our fossil fuel resources. The MHD concept in a somewhat different form can also be combined with an atomic power plant to achieve increased efficiency.

MHD is, however, still a concept rather than a reality. Developing materials for the channel capable of withstanding the 5,000° F. temperatures at which the fuel is burned poses a major technical problem. The low level of funding for the research—only some $2-5 million a year—has hampered progress. The MHD generator could be designed with a built-in system of air pollution control, and could be a means of allowing us to use all of our fossil fuels cleanly and more efficiently. It could save energy and money too, but it remains an option that exists only on paper. Technical problems may defy solution, but we won't find out at the present level of effort.

The fuel cell is an alternative technology for generating

electricity that has far-reaching implications for conservation and reshaping the energy industries. It is a "little black box" without any moving parts that chemically converts fuels into electricity. It is economical as a household power plant yet it comes in larger sizes. If hydrogen is the fuel, it operates at efficiencies as high as 60-70 percent. And the only discharge is water. If methane (gas) is used as the fuel, it discharges carbon dioxide, which is not a pollutant. Fuel cells are a spin off from the space program where they were used to generate electricity and water for the astronauts.

Natural gas or oil can be used in a fuel cell, but these are in short supply. Hydrogen is an ideal source of energy for the fuel cell, a source which like gas could be transported directly to consumers by an underground pipeline system. Hydrogen is scarce and expensive today. But if low cost methods of generating electricity in large plants at remote locations could be perfected using the breeder or solar, geothermal or fusion power (see Chapter 12), then hydrogen could be manufactured near the power plant by electrolysis of water (with valuable oxygen as a by-product). Hydrogen could then be transported economically by pipeline for long distances with little loss in energy. In fact, for distances in excess of 200 miles, hydrogen can be transported more economically than electricity, and the transportation system is underground all the way to the consumer.

Because a fuel cell could be located in the home, apartment building, shopping center or industry, it could provide electricity completely independent of the utility company. The fuel cell thus could not only conserve our fuel supply but also facilitate competition in the energy industry, providing an incentive for better service and reasonable rates. Imagine the change in bargaining power if you could tell your local electric utility to peddle its kilowatts somewhere else.[201]

The fuel cell could provide a real competitor to central station electric power production while generating electricity much more efficiently than large plants and with little or no resulting pollution. One would imagine that such a promising technology would be the subject of an intensive development program by a federal government concerned with the problems of energy and the environment. As it is, only a few enterprising gas and electric utilities are investing, and the research effort remains low-key. There is no federal support for fuel cell development. The natural gas industry has shown commendable initiative in proceeding with this long-term research program, but many companies began to lose interest as the shortage of gas became a reality. Electric power companies were invited into the program, which now has as its first commercial objective a fuel cell to use in an electric utility's substation.

Fuel cell research is proceeding at a lackadaisical pace. Unless the federal government takes an interest in perfecting the little black box, it's not likely to play a major role in saving fuel for America's future or for affording an alternative to central station power.

Putting the waste heat from electric power plants to useful work is a major incentive for developing smaller power plants that could be located in shopping centers or other urban locations. These so-called total energy systems have difficulty in competing with large power plants in the efficiency of electric power production alone. However, if we consider both electricity and waste heat as valuable products, on-site power plants become more attractive.

A breakthrough for using waste heat would occur if large central station plants could be made clean enough and safe enough to be located close to urban centers. The waste heat could be used instead of fuel as a source of commercial energy. If a plant could be built within a few miles of a new community, a central district heating system could be built instead of individual furnaces. District heating systems are in operation in Stockholm and other Swedish cities as well as in Poland and the Soviet Union. They make use of waste heat from power plants or in some instances a single large oil "furnace" which makes steam heat for a community at substantial savings over individual furnaces. The waste heat could be applied to the air-conditioning of buildings as well as heating them. "Waste" heat also can be used effectively in sewage disposal plants, in removing snow from city streets and for heating greenhouses that will increase production of valuable crops.

It is industry, however, that presents the larger and more practical opportunity for marketing the waste heat from power plants. The industrial use of process steam is about 17 percent of the nation's total energy consumption, which is roughly equal to the volume of heat energy potential wasted at all the power plants in the nation. In earlier decades the paper mills and other industries that use large quantities of process steam operated their own electric power plants on the same site and did indeed use the "waste" heat in their operations. More recently, however, the utilities have been able to sell electricity to large industrial customers at prices much lower than the cost of self-generation. Also, until recently, the low price at which natural gas and coal were available for making process steam gave the major users little incentive for complicating their siting problems by trying to locate next to a power plant.

The electric utilities are having great difficulty finding sites for their own plants. With rare exceptions they lack the incentive, responsibility or imagination to seek out industrial customers able to use waste heat which might locate adjacent to their plants. (An exception that points up the viability of the idea is the contract signed between Dow Chemical Company and the Consumers Power Company in 1967 under which Dow will buy steam as well as electricity from a nuclear facility to be built at Midland, Michigan.) Here is a situation which cries out for government leadership.

It is difficult to overstate the magnitude of the energy potential of the waste heat from power plants, today and in the future, and the benefits to society from putting it to use. By the year 2000 there is every reason to believe that 50 percent of total energy will be used for electric power production as contrasted with the current 25 percent. If we continue to "waste" two-thirds of this energy potential we will then have a waste heat potential that could satisfy one-third of our total energy needs. The waste heat can be cut back by the more efficient methods of generation discussed earlier. But it is also a potential source of clean energy since society has already paid the price in environmental degradation for the creation of this waste heat and will suffer even more pollution if it is dumped into the water or air.

Another way to save energy in the electric power sector is through new technology for power transmission lines. Most conventional high-voltage lines now lose from 5 to 30 percent of their load during transmission, depending largely on the distance the power is transmitted. But new technology in the form of superconductors and cryogenics can cut losses almost to nothing.

A superconductive system works by surrounding an ordinary transmission line with a super coolant, such as liquid helium or hydrogen, in a buried, insulated pipe. The temperature of the coolant reaches about 420° Fahrenheit below zero. Research has not only shown that line losses can be minimized, but also that the capacity of ordinary wires can be greatly increased. In 1972, General Electric staged an experiment with a short superconductive line which carried more than 3500 megawatts for several days with no damage to the line. This is seven times the load which can be carried by the best above-ground cables. An added advantage of superconductive lines is that they are buried underground and require only a small fraction of the land currently needed by overland transmission facilities. For example, a new addition to the Four Corners power plant will eventually need ten new parallel transmission lines to transport the added capacity to the West Coast. These lines will require more than 100,000 acres of land. The whole load, however, could be carried by a single underground superconductive cable in which the power line itself would have a diameter no larger than four inches.

Work is proceeding on these newer ways of transmitting power but progress is slow. Federal research support is very limited—only about $2 million a year. With proper effort and funding, these transmission lines could be developed for commercial use in a few years. Indeed, it is essential to have such lines if large plants are to be located in remote sites and tied together in an interconnected grid system for maximum efficiency, safety, and reliability.

CHAPTER 12

Tapping the Eternal Energy Supplies

In the past two chapters we have reviewed some of the many options open to us for both reducing our energy needs and increasing our energy supplies. With a determined effort in both directions we can ensure the viability of our high-energy society for at least a few more decades.

Yet as long as our primary sources of energy are fossil fuels, which are finite, we are in a race against time. Within a few decades these sources will peak out, and eventually they will run out. The nuclear breeder reactor holds promise of vastly enlarging the energy resource base. But we dare not face the future with this sole option, which is at best only a limited answer and one which to many represents more a threat than a promise.

The time we have in which to develop "something else" is uncertain. In a sense, the energy crunch has already begun. But the world can probably buy another 25-50 years through conservation and development of the resources discussed in the prior chapter. If the nuclear option proves too dangerous, however, the crunch will come much sooner—in fact immediately.

From one perspective there is already an acute energy crisis. Present fuel supplies could not support worldwide energy consumption at even half of the U.S. per-capita consumption. And many believe it is immoral for the U.S. to pursue wasteful and self-indulgent energy policies which assume that most of the world will remain energy-starved—immoral, as well as dangerous for the peace and security of the energy gluttons.

It is not only the fear of running short but also the environmental problems with all of the major sources previously discussed that compels us to seek "something else." Even conceding that new energy sources always "look cleaner in a test tube," they do hold promise of being much less damaging to the earth and its people.

Fortunately, there are forms of energy that cannot be depleted, that for our purposes are eternal, since they will last as long as the earth itself. Three such sources are known: geothermal energy, solar energy and the power available from controlled thermonuclear fusion. Solar and fusion power are virtually limitless in their potential for serving the energy needs of the world, provided we can overcome the immense technological barriers which now block their widespread use. In a sense we are surrounded by energy beamed down on us from the sun and rising toward the surface from the magma, or molten rock, that forms the earth's core. Herein lies the largest challenge we have yet faced in our scientific history. These sources of power are capable of supplying the energy needs of mankind literally forever. Can we marshal the technological, economic and political resources needed to harness them? And can we do so before existing energy sources run seriously short, or overwhelm the earth's environment?

A first step is to know about these energy forms and to generate the force of public awareness to compel government commitments to launching the programs and to spending the billions of dollars necessary for their development. The major difficulty is public indifference stemming largely from the failure of our leaders to give the future a high enough priority over the special interests favoring the status quo. As a nation we have a track record of failing to do today what is needed to avoid tomorrow's problems. Opposition to such commitments comes from the energy industry, which seeks available funds to perfect the coal resources and other products they are currently selling. Who will speak up for attempting to harness the sun? Surely not the petroleum industry or the companies that sell nuclear power plants. Thus our future needs must compete for attention and dollars with powerful vested interests presently providing energy supplies. We will deal with these policy problems in the next chapter. First let us see what science and technology might develop for us if given the opportunity.

GEOTHERMAL ENERGY

For a nation running short of energy it should be a source of comfort, and embarrassment, to realize that the earth itself is a gigantic storehouse of heat and energy that we have scarcely begun to tap. The earth's core is molten rock, or magma, and the intense heat radiates out toward the earth's surface. The outer layers of the earth's crust pretty well contain the heat, but it is close enough to the surface of the earth to be reached with modern well-drilling equipment. In fact in a few localities, some of this heat makes its way to the surface and is found in volcanoes, geysers and boiling springs. But these eruptions are merely tiny leaks in a great natural storehouse of energy.

The situation may be compared to our petroleum resources prior to extensive exploration and development. We know the geothermal energy resource is in the earth, and in a few places it has actually "seeped" through to the surface. But the extent of the resource and its cost of development are unknown. Unlike our intense curiosity about petroleum, however, the U.S. has not undertaken any extensive search for geothermal energy.

For many years, scientists have dreamed of tapping this vast source of heat to provide clean, cheap, and virtually perpetual energy. They envision geothermal power plants outside our large cities: geothermal heat plants on our coasts to desalinate seawater; even vast underground cities, heated and lighted by energy from inside the earth. But geothermal power remains largely untapped for the same reason as solar power and our other large, unused energy options. Fossil fuels have been available in sufficient quantities and at reasonable cost.

Geothermal energy has been used for some time on a small scale. In Boise, Idaho, homes have been heated by underground steam since 1890. A 30-megawatt plant using underground steam to generate electricity was built in northern Italy in 1904. Other geothermal plants around the world have a total generating capacity of about 1 million kilowatts, including an 82,500 kilowatt commercial electric power plant at "The Geysers" in northern California.

These existing commercial applications are small, but they reveal that geothermal energy is technically feasible. In some cases it was economical even during the era of low-cost abundant supplies of fossil fuels. Environmentalists who are impressed with the absence of air pollution and radiation dangers in geothermal energy now are being joined by others concerned with the shortage of energy in urging that we make a serious effort to find and use this power from the earth.

There are two basic categories of geothermal energy: (1) the prime areas where the heat is concentrated in reservoirs relatively near the surface and (2) the heat generally available at any location that is drilled deeply enough to obtain the desired concentration. The prime areas are in volcanic regions and sedimentary basins where the rock is sufficiently porous for the underground water to form a reservoir of heat as steam or hot water. This heat can be tapped and brought to the surface by drilling a well in a manner not unlike an oil or gas well. But unlike an oil well, geothermal heat tends to be replenished by the enormous hot center core of the earth.

A key question is how extensive is geothermal energy. Is it a small freak of nature or a major option for the future? The responsible U.S. federal officials discount the future contribution of geothermal energy. Typical is a conclusion of the Department of Interior that:

> Under favorable conditions, geothermal energy may be locally important to several areas in the Western states; however, it probably will be insignificant as a factor in national power capacity (less than 1 percent of total) through the year 2000.[202]

A more optimistic view of the U.S. geothermal power potential is advanced by scientists who have made specialized studies. They estimate the potential in the western U.S. at about 100,000 megawatts, and one estimate is as high as 10,000,000 megawatts.[203] These figures may be compared to the total U.S. electric power capacity in 1972 of about 400,000 megawatts.

Interestingly, the more optimistic views of the U.S. geothermal potential are supported by an independent study by a Task Group of the National Petroleum Council. After identifying potential heat sources in the earth beneath the U.S. that are the equivalent of more than 1 trillion barrels of oil and cautioning that we lack accurate information concerning that resource, the study concluded that:

> Nonetheless, it would seem that the heat that may be recovered from high-enthalpy masses of water in the deeper parts of sedimentary basins of the United States could be of considerable magnitude. It seems conceivable that it may overshadow even the overall total for oil. The petroleum industry, in particular, because of its familiarity with basins, seems to be in a strategic position to take advantage of these opportunities.[204]

Geothermal energy must be viewed as a worldwide source of energy, for reservoirs of steam and hot water are to be found beneath the earth's crust on most of the continents, a number of islands, and perhaps beneath the seas.[205]

Favorable conditions for developing geothermal energy exist in the nations of Central and South America, in Turkey, East Africa, in almost all the countries around the Mediterranean, in the Far East, along the "Circle of Fire" of volcanic action surrounding the Pacific, and in the Soviet Union.[206] Almost everywhere, the geothermal resource is, as in the U.S., an unknown quantity. However, the Soviet

Union has conducted extensive exploration for geothermal energy, and a Soviet scientist has stated that geothermal resources exceed all of the other vast energy resources of the Soviet Union.[207] Many of the developing nations are in areas of prime geothermal potential, and it is entirely possible that it may become the principal source of energy in those nations. Unlike nuclear reactors, geothermal energy can be developed economically in relatively small plants that could be used by a developing nation's electric power systems.

The question may well be asked that if "geyser power" is all that promising why hasn't it been developed more extensively? A rather candid answer for the U.S. has been supplied by the petroleum industry itself:

> The limiting factors in the development of geothermal energy seem to have been, not economics, but rather the novelty of the idea, a lack of sufficient pre-investment geological and geophysical evidence of the extent of the geothermal resources, and, until recently, the fact that a large fraction of the geothermal prospective lands could not be leased for that purpose.[208]

The report might have added that government statements tending to dismiss geothermal energy as an insignificant option encouraged this attitude of neglect and amount to self-fulfilling prophecies. At the moment, too, an optimistic picture of geothermal potential in the U.S. has little more to support it than the conventional wisdom that tends to write it off. The true extent of the prime geothermal resource is unknown because of limited development and inadequate information, a fact which the federal government confirms.[209] A federal leasing program now being launched may well result in development of some of the prime areas. But this will not reveal the actual magnitude of the resource or develop the new technology needed to bring the more abundant heat to the earth's surface.

A survey to determine the nation's geothermal resource base is badly needed. In the U.S. the prime geothermal areas are located in western volcanic regions on federally-owned land. Such a survey must not only locate all of the prime sources, the reservoirs of steam or hot water, but also conduct research and build demonstration projects to use the heat discovered. Otherwise, we cannot learn what we need to know about the size of the resource, its location, the cost of marketing the energy on a large scale, and the existence of possible environmental dangers.

Outside the prolific volcanic regions the temperature inside the earth's crust increases about 40° F. with each mile of depth. Thus by drilling a hole in the ground several miles deep at almost any point on the earth's crust, we can obtain heat of sufficient intensity to heat and cool homes, offices and factories, and perhaps even to be efficiently converted into mechanical energy. The economics are uncertain but not forbidding, if one considers the possibility of building new communities with district heating systems to avoid the expense of an individual furnace for each home. A few such installations are in operation in the Soviet Union and France.[210] We have not even conducted experiments.

Techniques have been suggested by scientists for extracting geothermal energy even in areas where there is no underground water, only hot rock. One method proposed by Dr. Morton C. Smith of the Los Alamos Laboratory would be to drill a well and pump water into it under sufficient pressure to crack the rock. A second well would be drilled alongside into the fractured rock thus permitting the heated water to return to earth by convection. The wells would be from 2-6 miles deep, which is within the range of conventional oil drilling rigs. Each well could provide a source of energy for a 100 megawatt electric power plant. Areas of "hot rocks" are to be found in the east in Pennsylvania and in upstate New York, for example, as well as in the West.[211]

Another possibility along similar lines is to use nuclear fracturing of large quantities of hot rock and then to extract the heat by piping water into the fractured area and creating steam which is conveyed directly to an electric power plant.[212] The cost estimates for this option appear competitive with existing power sources, but they are quite sensitive to the size of the atomic blast and are based on finding concentrations of heat at relatively shallow depths. There is also an inherent conflict between the cost assumptions and safety considerations. The dangers of triggering an earthquake rise with a larger blast in the volcanic areas in the West where the necessary concentrations of heat are most likely to be found. Even so, this technique might prove economically feasible in areas where the rock can be fractured safely by atomic blasts.

Firm estimates of the economies of large scale geothermal projects cannot be made without a survey of the resource and some drilling and experimental operations. Yet existing geothermal installations, while by no means typical, offer encouragement. Electric power is being generated at The Geysers in California at a total cost of 4 mills per kwh, less than half of the cost for nuclear power plants under construction. The comparison may be quite misleading because The Geysers is a prime source. Yet it does reveal that if the heat can be found in sufficient concentrations at relatively shallow depths, geothermal energy could be economic for electric power production.

Economic prospects for using geothermal heat directly in homes, offices and in industries also appear to be attractive, provided of course, that the city or industrial park is located near the geothermal energy. Cost data from prime areas already developed in Iceland, as well as Italy and New Zealand suggest that geothermal energy can be developed "at very low costs as compared to heat derived from fuels."[213] The recent doubling of oil prices provides a much broader economic horizon for geothermal energy.

The economics of geothermal energy could be enhanced if multi-purpose development proved feasible. The hot geothermal water from a wet field could be distilled to make fresh water without any external source of heat.[214] And water will be desperately needed in the Western states where the geothermal potential appears to be present. Also the possibility of extracting valuable minerals from the geothermal waters offers another by-product possibility that could further enhance the benefits of geothermal development. But all of this will remain largely a matter of conjecture until demonstration projects are built.

The major impetus for geothermal development however, is not lower costs but rather the shortage of relatively clean sources of energy. Environmentally its advantages are readily apparent—freedom from the air pollution caused by fossil fuels and the absence of the danger of radioactivity caused by atomic power. These are clear and very important benefits which provide a strong environmental incentive to pursue the geothermal option. But there may be offsetting environmental problems of unknown magnitude.

A troublesome concern is the possibility that withdrawing or reinjecting "hot water" might stimulate fault zones beneath the surface and cause subsidence or even earthquakes. The best geothermal prospects are in areas with a history of earthquakes. But there is no direct evidence that earthquakes are related to existing geothermal installations and experts differ on whether the danger is real. The present evidence is inconclusive and more research and monitoring are needed.

There are also uncertainties about the effects of gases contained in the geothermal steam and the disposal of the geothermal brines. But these problems appear to be susceptible to engineering solutions as part of a development program. Geothermal energy does not appear to pose the fundamental threats to the environment inherent in the use of fossil fuels and nuclear fission, assuming of course, that the earthquake danger is indeed overdrawn.

One environmental question will have to be answered according to local popular sentiment. It is the question of when a development opportunity should be passed up and a site left in its natural state. For even if all other environmental concerns are resolved, geothermal development inevitably means industrialization of an area because this form of energy must be used where it is found. The low-grade heat cannot be transported very far.

In facing the land-use issue we may agree that national parks are clearly off limits, but other areas of scenic beauty do not present such a clear-cut choice. A national land-use policy must be evolved and careful guidelines drawn so that we do not allow geothermal power development to preempt prime recreation areas. But we also should avoid creating a vacuum in the decision-making process and thereby ensuring that no development at all takes place. Most of the prime geothermal sites are located on federal land. The federal government could show the way by a national survey based on actual drilling that identifies the resource and classifies areas either for development or for recreational preservation as appropriate.

The U.S. has an important responsibility, both to its own people and to the developing nations of the world, to launch a more intensive program of exploration and commercial demonstration for geothermal energy. Such a program is urgently needed to remove the uncertainties surrounding this abundant, clean form of energy. Can it become commercially attractive, or is its seeming promise just an illusion? We will never know until there is a concerted effort to find out, a task which the government must undertake at once.

SOLAR ENERGY

The sun's rays are the ultimate source of life on this planet and it is an obvious source of commercial energy. Certainly solar energy is plentiful enough. The land area of the U.S. receives more than 100 times our total current energy consumption in the form of sunshine even if we were to use the solar energy that reaches us at only 5 percent efficiency.[215] The energy supply is self-renewing and in its natural form poses no threat to the environment of which it is the central part.

But the reasons we have not gone directly to the sun for our commercial energy are also readily apparent. On even the brightest and hottest day, solar energy is not concentrated enough to provide the power for modern-day commercial and industrial needs. And, of course, the sun doesn't shine with the same intensity all the time, and it does not shine at all on many days and at night. Moreover, the sun's availability differs widely among geographic areas.

Nevertheless, solar energy has intrigued individual scientists over the years, and many ingenious gadgets have been invented to put the sun's rays to limited commercial use. A solar-powered steam engine was a feature attraction at the Paris Exposition of 1878. More practical applications involve the use of mirrors and absorbent materials that collect and concentrate solar energy on a small scale to heat or evaporate water.[216]

Energy From Growing Things

Solar energy sustains all the living organisms on earth. In a sense every growing thing is a "power plant" converting solar energy into plant life through a process called photosynthesis. Only a century ago most of our commercial energy came from the solar energy stored in trees, which were cut and burned. It is not likely that we will go back to wood, but some scientists believe that crops that are more efficient in the conversion of solar energy could one day be grown for the specific purpose of providing fuel.[217]

Our forests and woodlands are needed for recreation and our growing demands for lumber and pulpwood. But there are more than 200 million acres of uncultivated land in the U.S. suitable for growing crops. Cereal grain could be grown and converted by fermentation directly to ethyl alcohol, which can be used as a motor fuel. If we used 100 million acres for this purpose we could replace 12 billion gallons of gasoline, more than 10 percent of our 1972 consumption.[218] Such plants as chlorella are relatively efficient converters of solar energy, and there may well be others yet undiscovered that would be exceptionally efficient. The problem with using cereal grains for gasoline or other crops for fuel is one of costs. While no firm data are available the costs appear to be higher than conventional sources.[219] Agricultural residues such as straws, shells, hulls, corncobs, etc., could supply about 4 percent of total U.S. energy demands—as much as all our hydroelectric power—if they could be economically collected and burned in power plants. The costs of collection are so high, however, that this form of energy would cost about twice the delivered cost of competing energy sources at current prices.[220]

As we move into an era of energy scarcity, a determined

search for the most efficient form of "flower power" should be a part of our research effort. Until now, there has been scant interest in finding plants that, because of their rapid growth, are highly efficient sources of stored solar energy. Such research would not only benefit the U.S. directly but might also greatly assist the developing countries with much uncultivated land, surplus labor and an urgent need for energy in small convenient packages.

Solar Heating—
An Available Potential

Solar energy may be our most neglected energy option. Yet despite our neglect we know enough today to be certain that solar energy can be harnessed economically to supply much of the space heating and cooling for new buildings. The opportunities for immediate solar energy applications are well known, but the so-called energy experts have scoffed at them in the past. Solar energy can no longer be laughed off. It is a source of immediate help in solving the clean energy shortage.

The technology is available to use solar energy to supply a sizable fraction of the hot water, space heating and air-conditioning requirements of homes in many parts of the nation. Indeed, experimental solar-heated homes are already in existence in Boston, Massachusetts, Washington, D.C., Denver, Colorado, and elsewhere, and they have a satisfactory record of performance. These techniques must be perfected for mass production, but no new inventions are needed.

The concept of a solar home is really quite simple. A portion of the roof of the house is used as a "collector" made of a black surface to help absorb the heat and filled with water or air covered by sheets of glass to provide a "greenhouse" effect. The heat is then transferred for storage into a large water tank or gravel bin and circulated throughout the house in much the same way as a conventional hot-water or air system. A supplemental heating system may be required for very cold days but the solar energy system supplies the greatest portion of the fuel supply, the portion varying, of course, with the region of the country. The same system also provides hot water.

Solar heat can also be used to cool a building. There is nothing novel in the idea—the heat is used to power an absorption refrigerator, the type which is usually operated by fuel. Experiments show that the concept is technologically promising. The economics also seem attractive because there is a high degree of correlation between the availability of solar energy and the need for air-conditioning. The same basic collector system installed for solar heating can provide air-conditioning as well. In fact, the cost of heating and cooling a house with solar energy as compared with conventional sources may be extremely competitive. A comprehensive economic analysis with electrically heated homes even before the latest escalation in fuel prices shows that the fuel savings from solar energy more than pay for the extra investment, in every section of the U.S. except the Pacific Northwest.[221]

A good example of the economy of a solar heated home is provided by a three-bedroom solar-heated home built twelve years ago in the Washington, D.C., area. The solar heating system was added at a cost of $1,500. Over the years the system has required less than $5 per year of supplemental fuel oil to keep the house warm. It would appear that the savings of some $200 annually have more than paid for the investment.[222]

Potentially even more competitive in all regions of the country is the solar home—a residence both heated and air-conditioned by solar energy. The solar home equipped with supplemental electric heat for very cold days can compete with conventional homes using oil and gas heat and electric air-conditioning.[223] Cooling a house with solar energy would relieve the electric companies of the air-conditioning loads that contribute so heavily to system peaks in most regions. And any electricity required to supplement the sun in the winter in most areas would be in the off-peak months and no additional electric generating capacity would be needed. Another application of solar energy that appears feasible and economically attractive is a solar-heated swimming pool.

The water is heated by flowing through a solar heater next to the pool. A model built in Northern California, for example, would pay for itself in three years.[224]

The prospects for economical solar homes seem promising but at the moment we only have paper studies to support this optimism. Aside from individual experiments there have been no recent demonstration projects to supply cost data and show people that solar energy will work. What is most needed is a demonstration of the integrated solar heating and cooling system that seems most attractive, and a few such buildings are being constructed under grants from the National Science Foundation.

The basic pieces of hardware needed to build a solar house are readily available.[225] But private builders have little incentive for proving them in practice. The solar home is initially more expensive to build than a conventional one, and consumers are largely ignorant of the potential savings on their energy bill. There is no incentive for the builders to innovate. On the contrary, the Federal Housing Administration and private lending institutions discourage solar homes because they base their loans solely on the initial cost of the house. Thus these institutions are an obstacle to innovation that costs more at the outset but that would save the homeowner money over the life of the house.

Solar energy homes are an attractive near-term possibility, but they will not be built without a concentrated effort by the government to make them a commercial reality. First, there must be a demonstration program in various parts of the country to prove out and perfect the hardware and integrate the solar system with supplemental heat sources. Once there is workable technology and accurate data on costs, consumers can be expected to exert their influence on builders. An even more powerful influence could be exerted by the Federal Housing Administration if its construction regulations and standards and terms for loans were revised to encourage the residential use of solar energy. If it can be shown that solar energy will save consumers money while conserving our resources and minimizing pollution, then surely the government—which dominates the housing industry through loan guarantees, subsidies, and tax incentives —can take action to encourage its use.[226]

The best potential market for rapid solar energy development is the residential sector. The technology is simple, and enough solar energy reaches the roof of an average house to meet heating and cooling needs. It would, of course, be difficult to fit existing homes with the system economically, but in many cases it could be done. In any event it would be decades before a sizable energy-saving contribution could be made. But even small percentages mean large savings in the U.S. energy market. By 1980 most new homes being built could well be solar heated and cooled, and the savings could at once be measured in millions of barrels of oil each year.

Large-Scale Solar Power

Solar energy is abundant for all of our needs and will last as long as life on the planet earth. The key to its large-scale use is to convert this diffuse and unreliable source into a concentrated, steady source. As a source of energy, the sun's rays are diffuse and intermittent. And in the past, the large capital costs and technical problems of capturing, concentrating and converting it to a reliable source made any serious effort to harness solar energy seem foolish. In a sense the industrialization and centralization of society, which followed the invention of the steam engine, is based on the availability of energy in a much more concentrated form than the natural rays of the sun. For example, a modern steam boiler requires heat 500 times as great as the average density of solar energy. If we could economically convert solar energy into electricity or some other concentrated form, then it could serve the apartment buildings, offices, motor vehicles and industrial establishments that make up the other 90 percent of the energy market, which requires energy in a greater concentration than that which the sun can deliver directly.

As it happens, in the course of the U.S. space program we have learned to convert the sun's rays into electricity with a fair degree of efficiency. A solar cell made of silicon, a material abundantly available on earth, has powered our unmanned space craft by converting the sun's rays into electricity. The problem is that these solar "power plants" cost about $250 per watt or about 500 times as much as power from a conventional power plant, even considering that the fuel is "free." And, of course, a solar cell would not produce power reliably here on earth.

Nevertheless, the cost picture is by no means bleak. These solar energy cells have been hand-crafted to meet stringent space requirements and are assembled much like a Swiss watch. There has been no concentrated effort to apply mass production techniques to reduce the cost. Solar enthusiasts believe that a research and development effort to improve efficiencies and develop a continuous silicon ribbon for mass production could reduce the cost a thousandfold, down to $.26 per watt.[227] Cadmium cells also apparently can be produced at about the same cost or possibly cheaper.[228]

This cost would be lower than the cost of a new nuclear plant, which approaches $.50 per watt and to which fuel costs must be added. But a solar cell would not produce power on a steady basis. Even so, if cells could be produced for $.26 per watt the supplemental power would be economical for many uses.

Solar cells made of silicon, or perhaps cadmium, thus hold promise for producing solar powered electricity here on earth at prices competitive with conventional systems. The first application, again, may be right on the roof of your house. Solar cells located at the point of consumption eliminate the cost of transmitting the electricity from a distant power plant. The cells can be competitive if they meet the delivered cost of electricity, which is more than twice the cost of generation and currently averages more than 2 cents per kilowatt hour. If the contemplated cost reductions can be achieved, solar cells could economically supply much of the electricity needed for a home.[229]

Arrays of solar cells may also prove economical in shopping centers, schools and other buildings with large roof areas. Furthermore, they could possibly be located on structures above existing blacktop areas such as parking lots.

Open uncultivated land could be used to harvest solar energy as a rich crop that would not disturb the terrain. And some industries that have large needs for off-peak power may find that the energy from solar cells, even if available only in the daytime, would be economically attractive.

The major breakthrough for solar energy, however, would be large, central station generation of electricity. Here the economics are most difficult and uncertain but the possibilities are exciting. Unfortunately these ideas exist only on paper.

One idea for research advanced by Aden and Marjorie Meinel, two well-known solar-energy advocates, is the use of large solar power stations in the desert areas of the Southwest. This area is ideal for solar energy because the climate is hot and the land is not well suited for other uses. Assuming that super-conducting extra-high voltage power lines are developed, solar energy would be transmitted to distant markets as electricity. Alternatively the electricity could be converted into hydrogen in the desert and the hydrogen piped throughout the system very economically. Hydrogen, as we have seen, is a pollution-free fuel suitable for almost any use, including powering an automobile.

A key to the Meinels' solar power plant and similar ideas by other researchers is the use of a sophisticated coating on the collector which was originally developed for the space program. The coating permits the sun's rays to pass through but traps the infrared radiation and does not permit it to escape. When applied to a solar energy collector, this coating traps the heat. A number of these solar collectors, basically similar to those on roofs of homes, would be located in the desert and the heat transferred to a central location in pipes coated with the same selective coating. The heat would be stored in a central storage tank and used to run a conventional steam power plant. The system thus would provide a reliable, continuous source of energy to drive a large, conventional power plant without conventional fuel supply or environmental problems.

This concept, which suggests a means of harnessing solar energy on a large scale, has not been tested even in a pilot plant. If it proved economical, a desert area of 115 square miles in Southern California and Arizona could produce 1,000,000 megawatts of capacity—three times this year's total electric power capacity of the nation.

This means, however, that a 1,000 megawatt solar plant in a prime desert location would require about 13 square miles for the heat collecting network as compared to less than 1 square mile for a nuclear or fossil-fueled plant. If such a plant were coal-fired, however, as are the new plants in the Four Corners Area in the Southwest, the area of land strip-mined over the life of the plant could be some 35 square miles. Thus the land requirements for solar power plants are really no greater than for the coal-fired plants. And the solar-collecting system only shades the desert while the strip-mining of coal ravages it.

In the absence of any demonstration projects, cost estimates for such large solar facilities are, of course, highly speculative. Paper studies suggest that electricity could be produced at the power plant for about 2 cents per kwh, which is about twice the comparable cost from a nuclear power plant being built today. This cost comparison is sufficiently close to warrant a major research and development effort, particularly in view of the uncertainties surrounding nuclear energy and the pollution-free, self-renewing nature of the solar option.

If solar power plants in the desert prove economical, the concept and technology could have widespread application throughout the world. The desert areas cover 14 percent of the earth's land area, some 7.7 million square miles. The largest are the Sahara in North Africa[230] and the Great Australian Desert. But other sizable desert areas are located in the Middle East, Mongolia, the U.S.S.R., Argentina, Chile and South Africa, as well as the United States.[231]

Possible sites for large scale use of solar energy are not confined to the world's deserts, although they alone have the potential to meet all the demands of a high-energy civilization. Another concept equally challenging but perhaps somewhat more problematical is to locate a solar power station in space. The power plant would consist of a large number of solar cells on a satellite placed in synchronous orbit around the earth's equator. The electricity produced by the solar cells would be used to generate microwaves, and these would be beamed some 22,000 miles directly to a receiving station on earth. When they reached the earth, the microwaves could be readily converted to direct current electricity and then fed into the conventional power grids of the utilities.

This concept may seem like Buck Rogers stuff (how many of those far-fetched ideas have come true?) but it is really not so esoteric. The space solar energy system is based on established scientific principles. No new inventions are needed. We know that such power can be generated in space, beamed to earth, and fed into commercial power grids. What we don't know for sure is whether such a system can be made reliable. And the greatest uncertainty is how much it would all cost.

The satellites would not be small. A 10,000 megawatt power station—enough to meet all of New York City's needs in the year 2000—would encompass about 25 square miles in area and weigh 5 million pounds. The receiving antenna would be 9 square miles.

One might well ask why even consider placing a solar power plant in space rather than here on earth. The extra costs and difficulties inherent in lifting the station into space and operating it there are bound to be very large. One reason is that the power plant would be "above the clouds" and could be a steady source of electricity 24 hours a day, 7 days a week.[232] Another is that the intensity of the sun's rays, the so-called solar flux, is seven times as great in space as the average on earth because the atmosphere surrounding this planet absorbs and reflects much of the solar energy.[233] Thus the same solar cells can produce much more electricity in space than on earth, and can do so around the clock.

The power plant in space would also virtually eliminate the waste heat or thermal pollution problem inherent in all types of central station power plants on earth, which waste over half the energy they consume. The microwaves could be beamed to earth and converted to electricity with efficiencies of approximately 90 percent, thus minimizing the waste heat problem.

All of these features make the solar power plant in space an attractive concept. But is it really practical or just "pie-in-the-sky"? No one can be sure but the concept certainly presents fewer technical obstacles than did going to the moon when the space program was launched by President Kennedy. In fact all of the basic technology for the solar power plant, microwave transmission and receiving station are in hand. And the space shuttle, now in early development, could provide the earth-to-orbit transportation system that would be necessary.

What is required to make the solar power plant concept technically feasible is an immense systems engineering effort.[234] An urgent effort would require more than a decade to develop the novel hardware required. But technical feasibility is within the range of possibility and would appear to involve fewer uncertainties at the moment than the development of fusion power, for example.

The question of costs—the economics of solar power in space—looms as a major hurdle. At the moment both the solar cells and the cost of boosting the cells into space make the concept prohibitively expensive. However, with the range of improvements envisioned through mass production of solar cells and the space shuttle, estimates by an independent panel are that electricity could be delivered on earth at 5 times current prices.[235] Proponents of the idea are naturally much more optimistic, believing that its costs could match conventional power sources. And they also make the valid point that if total costs to society were measured, satellite power plants might be a bargain even on the basis of conservative estimates. Even if the concept were pushed hard, solar power from space could not become commercial before the end of this century—but by then it could be a godsend.

No one can deny that the satellite concept is one of the most attractive ideas on the horizon because of its vast potential for meeting our twin goal of satisfying energy requirements and alleviating our environmental concerns. It is superior to any other concept in terms of minimizing side effects on the environment, being free of any polluting by-products and producing very little waste heat. Its land requirements are not unreasonable and require no destruction of the earth's surface.

Yet as with every source of energy there is a catch. The microwave beam from the space station to earth could be a lethal weapon if misdirected. Experts say the beam could be precisely controlled and its intensity at the earth would not exceed the density of solar radiation. Nevertheless, the possibility of an accident is a serious concern that should receive detailed investigation even though it does not appear to present an insurmountable problem.

The satellite power plant could well be the first really significant tangible benefit to mankind from our costly adventure in space. It would require a multi-billion dollar effort to perfect, and it is obviously a program that only government can be expected to undertake because of the uncertain, long-term nature of the return on investment. The satellite concept holds promise for all nations on earth because it is potentially a source of energy for all of them. Indeed the economical size of stations might require the participation of many nations. It is therefore an excellent project for international cooperation. Moreover, the reliance of many nations on a common source of energy would be a tangible way of bringing them together in peace and harmony and full appreciation of their interdependence. And it would reduce the danger of any single nation interfering with the satellites serving another.

A system of solar satellites placed in orbit as joint ventures by the nations of the world could relieve the anxiety over energy of "have" and "have-not" nations alike in the next century. A research and development effort to achieve such a system should capture the imagination of the United States and the Soviet Union, the two pioneers in space, but it should also involve every other nation on earth.

There are so many exciting possibilities for using solar energy that one wonders why they are not being pursued. An industry group gave this answer:

> Work such as this cannot be expected to be supported or motivated by profit incentives; it must be done for the benefit of all of the people of the world at their expense and as a result of long-range, visionary planning.[236]

People have been aware of the potential of solar energy, but thus far in the energy field the government has implemented "visionary planning" only for atomic energy. Ironically, in 1960, the Joint Committee on Atomic Energy held a hearing on the opportunities in solar energy. There were expressions of support and in an exchange with a witness, George O. G. Lof, even a surprisingly suggestive question by the Joint Committee's then Staff Director, James T. Ramey:

> MR. RAMEY. I have heard it often speculated if the Government had put $1 billion of the funds or the money it had put into atomic energy research and development into solar development, you might have a different picture on solar energy right now.
>
> DR. LOF. I am sure you would.[237]

But there was no follow-up. No appreciable funding has been provided because no executive agency has mapped out a program and requested the funds from the Congress. Thus 14 years later, in the midst of a shortage of clean energy, the situation is essentially unchanged. Solar energy is still our most neglected option awaiting a sponsor visionary enough to harness its vast potential.

NUCLEAR FUSION

The hydrogen bomb is the most awesome power on earth. Yet it is theoretically possible to control this mighty force with complete safety in a reactor that would fuel an electric power plant. The research program to tame the H-bomb could provide mankind with a super-abundant source of fairly clean energy in the next century. But it presents a scientific and engineering challenge that in many ways is much more difficult than placing a man on the moon or a solar energy system in space.

The thermonuclear reaction is much like the explosions

that occur in the sun. There is a joining together—a fusing—of the nuclei of light atoms such as hydrogen and the consequent release of tremendous quantities of energy. The dream of scientists since the 1930's has been to simulate and control this mode of energy generation in the sun and the stars.[238] But the pattern for atomic energy development has been first the sword and then the plowshare. Thus it was the successful work on the H-bomb in the early 1950's that gave impetus and encouragement to a concerted research program for controlled fusion power, begun under the name of Project Sherwood.

Fusion power is a very attractive energy research option because if it were successfully developed, the earth's uncertainties over fuel supply would be eliminated for millions of years if not forever. The source of fuel would be the deuterium and tritium found in and easily extracted from water. Through fusion, the energy of these atoms in one gallon of water would be equivalent to 300 gallons of gasoline. Thus, as former AEC Chairman Glenn T. Seaborg often pointed out, developing controlled fusion would be the equivalent of suddenly discovering that the Pacific Ocean basin was completely filled with a high grade fuel oil, instead of salt water, and that it would automatically refill itself 500 times as we consumed its contents.

The problems of perfecting fusion power are, however, about as mind-boggling as the fuel supply it would create. As one expert has declared, controlling the fusion process "is probably the most difficult technical task that has ever been attempted, bar none."[239] It is necessary, almost literally, to kindle a fire as hot as the sun here on earth and find some way to contain and sustain it as a long-term, reliable source of energy. Fusion power requires temperatures of about 100,000,000° C. There are no materials on earth that can withstand such heat. Everything vaporizes, and indeed it disintegrates even further into a mixture of electrons and nuclei that is really a fourth state of matter. It is not a gas, a liquid or a solid. We call it plasma.

The first problem is to build a fire of 100,000,000° C. In the H-bomb, an A-bomb is used to trigger off the reaction, but that is obviously out of the question for a power plant. The approach has been to "heat" hydrogen with a powerful jolt of electricity while confining the plasma thereby created. Since no material on earth can serve as a "bottle," the researchers have used magnetic fields to contain the plasmas. By ingenious arrangements of these "magnetic bottles" scientists have been able to confine the plasma in a stable condition and achieve higher and higher densities over a longer and longer period of time. The goal is to reach the combination of sufficient density over a long enough period of time to fuse the nuclei of the hydrogen atoms and produce more energy than required to start the reaction. Thus far the magic number has eluded the scientists.

The controlled fusion research effort was begun with high hopes in the early 1950's. In fact the prospects looked so bright that the United States, the Soviet Union and Great Britain each launched a separate effort in secrecy trying to be the first to tame the power of the H-bomb as a source of commercial energy. But word of the frustrations that each nation separately encountered leaked out. Finally in 1958, at the initiative of the Soviet Union, there was a wholesale declassification of all fusion research since no one had anything promising to hide. The research has been open ever since to the benefit of all concerned.

During the 1950-58 period each nation eagerly assembled ingenious gadgets confident that success was only a matter of finding the right shape for their magnetic bottles. The atomic physicists had been so successful in creating weapons that no one doubted "the technological omnipotence of physical thinking."[240] It was only when the nations compared notes that the need for much more extensive fundamental plasma research was recognized. If the fundamental lack of understanding of the behavior of plasmas had been recognized at the outset, the research programs might never have been started.

The return to the drawing boards appears to be paying off. In the past few years scientists have steadily approached the magic number of density of hot plasma over sufficient time to prove the scientific feasibility of controlled fusion. And it has been the Russian scientists who have led the way.[241] The Russian effort, measured by manpower, is three times larger than the U.S. program, and their results reflect the difference. But generally speaking, fusion research, unlike the other U.S. energy options, has not suffered from lack of funds. The U.S. alone has invested some $500 million since 1950 and the worldwide investment in this research option approaches $1 billion. There is a strong mood of optimism among scientists throughout the world that they will demonstrate the scientific feasibility of controlled fusion in the decade of the 1970's. This optimism is backed by results of recent experiments and the confidence derived from a theoretical basis for their machines, rather than the trial and error approach of the 1950's.

This optimism is reinforced by the emergence of the laser beam. It offers a radically different approach to fusion power which may by-pass the difficult struggle to contain hot plasma with magnetic fields. The laser beam shows promise of creating fusion power that is safe and plentiful in a fairly simple, direct way. A laser is a device for generating a very powerful beam of light that can be focused on a spot no larger than a few hundred-millionths of an inch.[242] In this new approach to fusion a high-powered laser is the source of heat. The fusion fuel is a tiny pellet, smaller than a pea, which is dropped into a vacuum chamber. Then the laser light is focused on it. The pellet is compressed to tremendously high densities and becomes so hot that fusion takes place and the pellet is vaporized into a plasma. What actually occurs is a mini-explosion with no need to confine the plasma at all. The process is repeated with a large number of pellets and a pulsating laser. There is no danger of a dangerous runaway explosion. The large number of mini-explosions (or "puffs") with a powerful, pulsating laser beam are each separate and there is no chain reaction. But they can create a large amount of heat energy which can be extracted from the vacuum chamber and used to generate electricity. In theory a large electric power plant of 1,000 megawatts could be fueled by 10 vacuum chambers burning 100 pellets per second.[243]

The experiments have not yet demonstrated that more heat can be extracted from the chamber than is required to power the laser. In other words we are not yet sure this system can be a net producer of power. But the scientists are quite optimistic. The current research is concentrating on the size of the pellet and on creating more powerful lasers.

If the views of those directly involved are taken seriously, the laser option, combined with the progress with the magnetic bottles, provides a strong basis for believing in the possibility of an early scientific breakthrough for fusion power. The research work of the Atomic Energy Commission on both options, and especially in the laser work, has been greatly expanded.[244] The funding for the U.S. fusion program is now at a level of some $100 million per year and even larger efforts are underway abroad. A private company is also pursuing an independent laser fusion research program with venture capital.[245] After almost a quarter century, scientific feasibility has yet to be achieved by either the government or private enterprise, but scientists involved say it could happen by 1975.

In assessing the prospects of fusion power we must remember that achieving a breakthrough in the laboratory is only the first step, albeit a giant one, on the long road toward commercial power. It took some 30 years and $3 billion to perfect current nuclear power plants after the first demonstration of the fission process at Stagg Field in Chicago in 1942.[246] Scientific demonstration of fusion in a sense will take us much further along because it also will provide some of the technology needed for the commercial reactor. But there is still the major engineering job of actually building a reactor. A nightmarish problem facing the engineers is to perfect a fusion reactor that can withstand internal bombardment from the very high-speed particles the fusion reactor will emit. Materials must be devised that can withstand a kind of stress to which no existing commercial material has ever been subjected.[247]

Interestingly, the problems of building a reliable commercial system are more straightforward with the laser option than with the conventional approach to fusion, primarily because the confinement of fusion with magnetic fields is abandoned. However, the materials problem is, if anything, more severe.

The task of building a fusion reactor has not even begun. It could be as elusive and frustrating to the engineers as proving scientific feasibility has been for the scientists. But the difficulties ahead are no reason to slacken our long quest for fusion power. The potential benefits justify as determined an effort as we can mount.[248]

Fusion appears to offer a complete success on two vital counts. Fuel supply would become virtually unlimited and there would be no safety problem as with current nuclear plants or the breeder. A fusion reactor is accident-proof because it would contain so little fuel at any one time. No more than a second's worth of reacting fuel would be in the reactor at any given moment, and the safety concerns arising from current nuclear plants and the breeder would be non-existent.[249]

The inherent safety of fusion power has lead some enthusiasts to suggest that the plants would be safe enough to locate in the middle of cities. Urban siting would indeed offer enormous economic potential not only in saving transmission costs but also in using the "waste heat" from the power plant for heating and cooling homes and commercial buildings with a consequent alleviation of air pollution as well. But it is by no means clear that fusion power will be entirely free of problems which could affect people. A fusion plant, for one thing, would not be altogether clean. Radioactive tritium would be produced, and while it is not as lethal as plutonium and other fission products, it is still a dangerous radioactive material. Presumably it could be recovered and used as a fuel in the reactor but tritium can diffuse even through stainless steel.[250] Thus fusion power is likely to raise its own special environmental concerns, even though the fusion enthusiasts suggest they will be minimal.[251]

Here again, the laser option offers at least theoretical hope of being superior. It has been suggested that pellets of pure deuterium be used as a fusion fuel with the tritium produced serving as a catalyst to keep the tiny fusion explosions going. This would drastically reduce the tritium inventory in the plant "to almost vanishingly small levels" and thus remove "the last possible objection to urban siting."[252] Moreover, the laser option would be economical in much smaller units than the other options. This would both facilitate urban siting and make fusion power more readily adaptable to the needs of developing nations.

The possibility of technical success raises another key uncertainty about fusion power—will it be economical? The assumption that fusion power will be cheap because the fuel itself is so plentiful and inexpensive ignores the question of the cost to build a fusion power plant. Since the fusion reactor merely replaces the boiler, except in the most ambitious options,[253] the plant itself cannot be built for much less than conventional plants. Fusion power will not be "free." The question is: Will it be competitive?

The consensus is that if the technical problems can be solved fusion power will prove competitive. There is more doubt on this score about the costs of laser fusion because the lasers themselves are expensive and not durable.[254] Some rough estimates suggest that the more expensive materials required will be offset by lower fuel costs. These estimates are supported by the great concentration of energy in a fusion plant which means fuel costs will be near zero. Thus even with high cost materials a fusion power plant should be capable of producing electricity at a reasonably low cost per kilowatt.[255]

After scientific feasibility if proven, perfecting a fusion power plant will require expenditures of hundreds of millions of dollars annually for decades. The steps to be taken are large scale experiments, demonstration plants and finally commercial scale production. If fusion power with all of its promise is to become a reality, the federal government must face up to financial requirements totaling billions of dollars. With a good, urgent program and excellent luck, fusion power could enter the commercial mainstream by the turn of the 21st century.

PART IV

TOWARD A NATIONAL ENERGY POLICY

CHAPTER 13

The Next Ten Years

Fear of energy shortages now pervades America. People have run out of gas and heating oil, lacked fuel to dry their crops and have been laid off because of shortages in the factory. The central question of energy policy is where our fear of shortages will take us. If the Arab boycott of 1973-74 is considered merely an unpleasant episode to be quickly forgotten as we continue our big car/glasshouse/plugged-in way of life, then our fears may lead to panicky retreat on many key questions. Environmental protection may lose out. Foreign policy may be shaped around appeasing Middle East oil producers. If we equate rapidly growing consumption of energy with the good life, then our other values must give way.

The next ten years—the period through 1984—probably will be decisive for our high-energy civilization. In this brief span, we can choose to persist in wasteful business as usual, although with increasing difficulty and under worsening pressures of shortages, until we are forced to cut back as the result of a severe energy recession. Or we can choose to look toward the longer term now and restrain our energy demand

and reorder our energy supplies so that our civilization and economy continue on a sustainable basis of energy consumption.

We need to channel the new public interest in the energy problem toward actions consistent with the values most Americans share. First and most important, we must overcome the fears growing out of current energy shortages. Americans need to be assured that sufficient energy will be available for essential uses. For the next year or two we will of necessity be changing our pattern of energy consumption to incorporate a policy of true conservation. As we make this transition, hopefully we can rid ourselves of rationing and other undesirable emergency measures designed arbitrarily to cut energy consumption. Rationing is a byproduct of the promotional spendthrift policies of the past. True conservation means freely chosen measures to save energy through greater efficiency.

Over the next ten years, we will need to expand our supply of energy because even if conservation measures succeed, some additional supply will be necessary to sustain our economic growth. Quite apart from the oil from the Arab nations—whose uncertain availability poses such a cruel foreign policy dilemma—increased supplies will be hard to find initially. As we shall see, domestic energy supplies can be enlarged somewhat in the mid-1970's and further increased in the years thereafter, but it will not be easy and there will be an environmental and safety price tag. The best prospects in the next year or two are to expand domestic coal production and produce additional oil from already producing reservoirs. The more we can save on the consumption side, the easier it will be to balance our energy budget.

A POLICY OF ENERGY CONSERVATION

A policy of energy thrift will yield double benefits. Obviously by using less, the consumer will have a lower monthly energy bill. Equally important, if the rate of growth in demand can be slowed, then price increases can be reduced. There are at least two reasons. One is that new energy sources are higher in cost than current supplies. Price increases are caused by the need to drill new oil wells far off shore, to import expensive gas from Algeria, to build new power plants and refineries that cost twice as much as existing plants, and to borrow more investment capital at high rates of interest. Another reason is that slower growth will remove some of the leverage that sellers can now exert. Prices reflect supply and demand. Slower growth will enhance the consumer's ability to buy at a more favorable price. Energy conservation thus means consumer protection.

A national policy of frugality and conservation in the consumption of energy should include these elements:

- Adoption of a new federal pricing policy for energy to reflect total costs to society and thus encourage efficiency in consumption.
- New tax laws to provide greater incentives for energy conservation.
- A "Truth-in-Energy" law to inform consumers of energy costs and opportunities for saving in the purchase of energy-intensive items such as automobiles.
- A research and development effort to perfect more efficient technology for energy conversion and consumption.
- Federal regulations and action programs for residential housing, commercial buildings, motor vehicles and perhaps other significant items to minimize energy consumption to the maximum extent economically feasible.
- Government investments in new communities, public transportation, and other activities—even building bicycle paths—that will save energy.

The first plank in a conservation platform is to wipe out the promotional policies now on the statute books, and replace them with a mandate for energy conservation. The promotional policies are engrained in the habits and attitudes of the energy industries and the consumers they serve. Repealing these outmoded laws and declaring energy conservation as an overriding national goal will not in itself change the lavish ways in which energy is used; it would, of course, be only a beginning. But it would signal our determination to forge new affirmative policies and programs.

A Conservation Pricing Policy

A new pricing policy will be at once the most effective and least disruptive means of encouraging conservation of energy. As we have seen, the cost of energy will increase to some extent in the coming decade. Federal government policy should ensure that price increases reflecting higher costs are designed to conserve energy as well as to protect the more vulnerable consumers. Fixing rates to conserve energy will mean imposing most of the increases on the larger volume consumers who are thus given strong incentives for the kind of energy-conserving discussed in Chapter 8. In that way, "promotional" rates will soon promote conservation rather than consumption. Such a policy will also alleviate the financial hardship of the small consumers.

Promotional rates that give large discounts to large consumers have most often been associated with the sale of electricity. These rates need to be changed. Most sales of natural gas to industry, however, are even more promotional and wasteful. Industrial sales of natural gas are typically priced only a few cents higher than the cost of gas in the field and reflect the very low marginal cost of using a pipeline that would otherwise be half-empty. More than 50 percent of the sales of natural gas are made to industry, and promotional pricing has been largely responsible. Residential and smaller commercial customers pay a much higher price, reflecting almost all of the overhead costs of the pipelines to transport gas hundreds of miles from field to market.

Large industrial customers, including electric power plants, could switch from gas to coal, after being given sufficient lead-time to install air pollution control equipment. Such controls are impractical for small users, such as households. Large industries switched from coal to natural gas in recent years primarily because gas was much cheaper. The Federal Power Commission should be given authority to fix the rates for sale of natural gas to industry, and it should be required to do so on a basis designed to discourage industrial use of natural gas. At the same time, such pricing would relieve homeowners and other small consumers in urban areas of the burden of much higher rates they would otherwise be charged. Higher prices for industrial sales of natural gas also would provide a powerful new incentive for greater efficiency in industrial consumption.

There are numerous other pricing techniques that encourage conservation of energy. One is to fix rates that more clearly reflect the costs to particular classes of customers. The major share of the retail cost of electricity is in capital needed to build power plants and equipment, costs that are present whether or not the capacity is used. Thus an appliance like a refrigerator using electricity around the clock should be charged a lower rate per kilowatt hour than an air conditioner used only about 20 percent of the time, the reason being that the capital costs are the same for both loads and thus the capital costs per kilowatt hour are five times as high for the air conditioner as the refrigerator. Yet current rates charge both loads the same rate. In a sense middle and upper income citizens with fully air-conditioned homes are being subsidized by consumers generally, and the subsidized rates encourage wasting electricity because homes are overcooled and under-insulated.

Air conditioning and other loads that occur in the summer peaks on most power systems are the root cause of the plant expansions that are now occurring. Charging higher rates to all consumers at the time of peak requirements would be a powerful incentive to reduce those peaks and thus reduce the number of electric power plants and the capacity of natural gas pipeline systems. Individual state public service commissions regulate the price of most retail sales of electricity and natural gas. A new declaration of federal pricing policy could provide specific marching orders for the regulatory agencies, requiring them to design utility rates to achieve these conservation and consumer protection objectives.[256]

Taxes To Conserve Energy

A new pricing policy must do more than accurately reflect the costs currently paid by the energy companies. Prices must also include the costs associated with using energy that society as a whole absorbs—the costs which are not paid by the energy companies or included in the price consumers pay. Some of these costs are gradually being brought into the price of energy through requirements imposed on the energy companies to prevent damage to the environment. Thus laws that require the restoration of strip-mined land and the control of air and water pollution, if they ever became fully effective, would remove the subsidy that society as a whole now provides to energy consumers. But while pollution continues, so does the subsidy. As long as energy prices are lower than *real* costs they promote a more rapid growth in energy consumption and contribute to the shortage of energy.

Energy conservation requires a policy of pricing energy on the basis of the *total* cost to society so that market forces can help reduce energy use to a level reflecting its real value to the consumer who pays for it. Pollution taxes can reflect these real costs and add them to the price of energy. The same taxes would perform double duty by offering a greater incentive for the consumer to use energy more frugally and also offering a financial incentive for energy producers to control pollution and thus avoid the tax. Such taxes are self-repealing in the sense that by spurring the development of clean energy systems they hasten the day there is no pollution to tax.

The pollution tax concept requires federal legislation but there has been no really serious effort to enact it. There have been some piecemeal proposals to legislate pollution taxes on emissions of lead and sulphur in recent years, but they received little backing even from the administration that proposed them. A bolder approach is needed. Legislation should authorize a tax on pollutants and provide a flexible mechanism for quantifying the tax and efficiently incorporating it into the pricing structure. An attempt to identify the pollutants in the law and fix the amount of the pollution tax, or fee, for each pollutant is too inflexible. Adjustments will be necessary as experience reveals how much damage particular effluents are causing. Such questions as which pollutants should be taxed, how much, and whether the tax should be leveled on the producer or purchaser of the fuel will require expert study. Congress should enact legislation which authorizes the imposition of such fees by an executive agency on the basis of substantial evidence after a public hearing affording all interested parties an opportunity to present their views. This should be entrusted to the Environmental Protection Agency, which has expert knowledge concerning pollutants and can formulate and administer pollution taxes to supplement and harmonize with the regulatory standards it imposes. It could well impose pollution taxes only to the extent that regulatory standards for clean air and water were not satisfied. In that way the tax system would not duplicate the regulations but would serve as a powerful incentive to assure swifter compliance.

Industry is capable of responding to higher energy prices by making capital investments to improve the energy efficiency of production lines. Therefore, a part of an energy conservation policy should be a general excise tax of 5 or 10 percent on all sales of fuel and electricity to industries above a certain minimum level for small business. The revenues should be earmarked to fund the research and development of more efficient energy consuming technology as well as to develop new sources of energy. Such a tax would help save energy and also provide needed funding for large scale research and development.

Consumers served by public power systems do not pay a fair share of the overhead cost of the federal government in the rates they pay for electricity. Additional taxes or payments in lieu of taxes should be imposed on these consumers as a matter of equity as well as to encourage conservation. The rates for electricity charged in the area served by the Tennessee Valley Authority, in the Pacific Northwest and in cities served by municipally or cooperatively owned systems do not contain any payments in lieu of federal income taxes. This income tax exemption, plus the lower interest rate paid on tax exempt government bonds, can lower the costs of electricity to the consumers of public power by one-fifth. A policy of pricing energy on the basis of total cost should, therefore, include a requirement that all utilities and the consumers they serve pay a fair share of the cost of government in the price of the energy they consume. Legislation should be enacted to impose a federal excise tax on sales by all utilities—public as well as private—roughly to approximate the average federal income tax component in the price charged by private utilities with an offset for taxes actually paid.

The basic purpose of these taxes to price energy on the basis of total costs to society is quite simple. In the final analysis there is probably no stronger incentive for consumers to use energy frugally than to charge what it really costs and thereby make it painful for them to continue wasting energy. The pocketbook approach to energy conservation is bound to have an effect, and so long as the money goes to the Treasury, and not the energy companies, the general public is better off.

There is one major difficulty with these suggestions for taxes to implement a policy of pricing energy on the basis of full cost to society. These taxes increase the price of a necessity such as energy on an equal percentage basis for all consumers, regardless of income. Lower income citizens would bear a disproportionate part of the increased tax burden.

As we have seen, energy is a sizable item of necessity in the budget of lower income families, and their financial problem is exacerbated by rising energy prices. Furthermore, poor people cannot cut back very much on their already limited consumption even if energy prices rise. It is the more affluent residential consumer who could be motivated by higher prices to insulate his home, to shop for the most efficient air-conditioning system, to buy a smaller car and to drive less. And the same is even more true for commercial establishments and industry where most of the nation's energy is consumed.

As part of the laws enacting pollution and other taxes to price energy on a full cost basis, there should be an energy credit for lower income groups. People earning $7,500 a year or less should be entitled to a complete refund of all energy taxes paid each year based on the average paid by a typical family, a refund that could be made available to all taxpayers as a flat tax credit. For those who do not pay income taxes the credit could be handled by the Social Security and Welfare Administration.

A Truth-in-Energy Law

Many consumers are now becoming energy conscious for the first time in their lives. But they need information if they are to make intelligent energy-saving decisions in today's marketplace.

A "Truth-in-Energy" law is needed that would require manufacturers of items using any appreciable quantity of energy to label their products accordingly. Consumers could estimate how much the item would add to their energy bills each year. The label would state the estimated annual energy consumption of an item, and the estimated annual energy bill to the consumer. Armed with this information, consumers could begin to ask questions and shop for items that would save energy and money when they bought a home, a car, or an air conditioner. For example, one brand of air conditioner might cost $50 more than another, but if the former saved $25 a year on the electric bill it would obviously be a better bargain.

The Truth-in-Energy law should go much further than mere labeling. It should require an "energy impact statement" as part of the plans for every sizable new commercial building or industrial plant. The statement, based on an analysis of the energy requirements for the structure, would report on the use made of all economically feasible equipment and designs for energy conservation. Guidelines for energy impact statements should be laid down by the federal agency responsible for energy conservation. Such a statement would encourage the use of heat pumps and other large energy savers, and it would discourage excessive lighting and ventilation (see Chapter 10). The impact statements would be available to interested government agencies and citizens, including prospective tenants of buildings whose rents would reflect the builder's success or failure in conserving energy. In a time of energy shortage, the requirement that competent engineers, architects and conservation experts be employed to ensure that all economically feasible steps are taken to cut down energy consumption would go far toward stimulating the marketplace to realize those savings.

Research and Development for Conservation

Research and development is usually considered solely as a means of developing new sources of energy. But science and technology also can develop new ways of using energy more efficiently. The government has not yet begun to formulate such a program. One promising area would be to engineer a car with vastly improved mileage efficiency (30-40 miles per gallon of gasoline) or one that could run on electricity at least for city travel. Many other energy-saving ideas need to be developed, such as new designs for buildings, energy-saving manufacturing techniques, improving the durability of manufactured goods and perfecting the methods of recycling metals (see Chapters 10 and 11).

Federal Action Programs to Conserve Energy

The public interest in saving energy is too great for halfway measures. Where important savings are clearly economical and attainable, government action should be taken to make certain these savings are realized. New buildings and automobiles are only two examples. A homeowner or a tenant in a commercial building may never have a chance to influence decisions in building construction. Despite educational programs, the owner may remain unwilling to spend an extra nickel on the original cost of the building to save money over the years on energy bills. No amount of consumer education, or level of taxes, is likely to stop consumers with plenty of money from buying large, gas-guzzling cars if they are still being built. Thus there is a significant role for direct government action in a few important areas where large energy savings can be achieved without inhibiting innovation in the marketplace. For new housing, the means to require that savings are realized is already in existence—the Federal Housing Administration's building code (described in Chapter 10). For existing buildings a large-scale federal loan program is needed to encourage owners to install insulation and storm windows and make other energy-saving improvements.

The automobile presents a real dilemma for public policy makers. Obviously it would be fine if the nation could count on Detroit swiftly to start making all of their cars smaller so that by 1985 we could achieve the goal of doubling the mileage efficiency of the average car on the road, a goal that is possible without eliminating the family car. Unfortunately, the performance record of the automobile industry offers us little assurance. It took federal legislation to force them to make cars that alleviated public concerns about automobile safety and pollution. On the question of mileage the situation is somewhat different because the motoring public is sending Detroit the message through their purchases in the dealers' showrooms. Many people are saying "bigger is no longer better," and saying it with their dollars.

Even so, there is no assurance Detroit will give mileage efficiency the high priority it deserves. The automobile industry has great incentives to try to maximize profits by going slow on retooling and continuing to promote sales of gas-eaters, especially if corporate executives feel bound to justify their failure to anticipate the gasoline shortage. Foreign auto-makers may apply too much competitive pressure for the U.S. industry to take this tack, but the automobile is simply too important an energy consumer to be left to the uncertain policies of the auto industry.

The approach that would leave the industry with the greatest engineering flexibility and yet provide the greatest assurance of success would be for the Congress to prescribe a gradually escalating minimum miles per gallon for new cars each year, beginning with the 1976 models. A minimum mileage, about 15 miles per gallon, applicable to about 50-75 percent of each manufacturer's output the first year could be met with compacts and still would permit some larger cars to be built until their mileage could be upgraded through better design. Both the minimum mileage and the percentage would then be increased each year on the basis of technical feasibility to achieve the goal of doubling the miles per gallon of cars on the U.S. roads a decade hence without eliminating a family-sized car. Doubling the mileage of all cars by 1985 would save one billion barrels of oil annually. It would mean the U.S. could keep rolling with 25 percent less gasoline in 1985 than we used in 1972, even with half again as many cars.

Related Federal Programs

The U.S. is fortunate that improvements in transportation systems, patterns of suburban and urban growth, and disposal of wastes can also save energy and help solve pollution problems as well.

Building automobiles with improved mileage efficiency is the backbone of an energy-saving program. But it is only the backbone. As we saw in Chapter 10, investment in mass transit is a vital part of an energy conservation policy. Using rail transportation for long-haul freight service and short-haul passenger travel between cities could cut in half the energy now consumed by trucks and automobiles on the highways. A fast train requires only one-eighth the energy per passenger mile of an airplane. The details of a program to revamp the nation's railroads are beyond the scope of this study, but a federal investment in an electrified rail transportation system linking the major cities of the U.S. is entirely feasible and offers a positive means of saving energy and money. Anyone who has traveled in the "bullet trains" in Japan knows that a 300-mile rail trip in two hours can be preferable to flying short distances. Anyone who has studied transportation economics knows that for a long-haul of more than 300 miles, rail transportation can be less costly than using a truck. With fuel costs rising, the potential savings offered by the railroad can be even greater.

What is needed is a national program to modernize the railroads. This will take time and money—billions of dollars. The federal government must take the lead, but with sufficient financial credit the U.S. railroads could regain their place as the spine of our transportation system. Surely any serious long-term program of energy conservation will be incomplete without building a modern electrified rail transport system connecting U.S. population centers.

The railroads suffer from more than neglect of tracks and rolling stock. The massive subsidy of their competitors through federal funding of highways and airports is a major obstacle to their revival. A program of energy conservation must therefore go beyond federal financial support for building new electrified railroads. It is time to abolish the Highway Trust Fund and the outmoded policies it enshrines. A program to encourage mass transit and fast trains should be substituted for the building of unnecessary new highways.

A national policy of energy conservation must also come to grips with the question of where new housing is to be built. It is senseless to speak of saving energy while permitting cities to expand in circles that each year become larger and larger, thus demanding more and more miles of automobile travel. New communities should be built combining housing, open spaces, and commercial centers, with public transportation systems incorporated in the design. The most important energy saving of all would be that people could again live close to where they can work, shop and loaf. Automobiles could be used mainly for pleasure again. In a sense the joyride would be resumed.

Conservation Policy—A Summary

A national policy of conserving energy would constitute a sharp break with the past for the U.S. But such a policy is no longer only an option, it is a necessity. The emergency measures of 1973-74 are not true conservation measures. We are forced to curb demand by government fiat because we have failed to conserve. A true conservation policy will satisfy consumer demands within the limits of our capacity to produce energy without self-destructive pollution and without mortgaging our foreign policy to other nations.

The actions recommended here can make a substantial contribution to avoiding serious energy shortages in the next ten years. The increase in the demand for energy in the next decade can be reduced to about two percent a year, as compared to about 4 percent in the past decade. Growth in demand for oil can be reduced even more and perhaps halted entirely as the growth is supplied by domestic coal and nuclear power. Such a reduction in demand would make an enormous difference. It could transform a potential national crisis into many small and manageable problems.

ENVIRONMENTAL PROTECTION—WHERE DO WE DRAW THE LINE?

A backlash has begun against the environmental movement. The retreat has not yet become a rout, but even a slow across-the-board retreat means that the problems discussed in Chapter 4 will become worse. Where do we draw the line?

The nation must move ahead to implement the clean air program or risk serious health hazards. There is mounting evidence that the standards are inadequate and need to be tightened to deal with the small particles invisible to the eye that results from burning coal or oil. The latest evidence on the formation of sulphates in the air suggests that sulphur oxide standards also may need to be tightened rather than watered down and delayed. The implementation plans can provide for sufficient time to build pollution control equipment for power plants and other stationary sources. But the "energy crisis" is no reason for delay. Reducing the environmental standards will not significantly enlarge energy supplies in short-term. The system of pollution taxes described earlier should be swiftly enacted to help speed compliance. By taking the profit out of pollution, these taxes would provide a monetary incentive for the energy companies to use pollution-control technology.

In contrast to air pollution, some of the water pollution caused by the energy industry seems far less deadly. For example, the damage caused by waste heat at electric power plants is far more speculative and remote from man. On large bodies of water, the huge and costly cooling towers may be doing more harm to the air and the landscape than they benefit the water. Thermal pollution is not in the same category as air pollution in deciding where to resist industry pressure for relaxed standards to facilitate more production of energy.

Water pollution from oil spills does pose a threat to the oceans and the beaches they touch. Efforts to control routine spills from tankers and prevent collisions must be stepped up. There is no conflict between these modest goals and increased energy supply. More troublesome is the dilemma posed by the oil at the bottom of the ocean within "spilling distance" of the nation's seashores, which are irreplaceable recreation areas. Yet simply passing up this "home-grown" oil and natural gas is an equally hard choice for a nation trying to balance its energy budget. We could save the coastal zones and still make use of any oil and natural gas that may be available off the Atlantic and Pacific shores by having the federal government, which owns the resources, explore for the petroleum itself. If it strikes oil, the government could use it to develop a strategic reserve for emergencies. The oil and gas would be shut in and as the quantities found justified it, pipeline connections would be built to existing refineries and natural gas pipeline systems.

Such an approach would be a compromise in the public interest between ignoring what may be a valuable and relatively clean source of energy and going all-out for a drilling program that threatens to destroy the natural setting of the coastal zones. In proving the undersea oil fields, the danger of spills would be minimal. Yet the petroleum supplies would be ready to flow quickly in any future emergency.

This is not to suggest that all offshore development should cease. On the contrary the federal government should move ahead to find out what the American people own off the Atlantic and Pacific Coasts and in Alaska, and to develop strategic reserves. Furthermore the Gulf Coast region off the Texas and Louisiana shores (but not Florida) have in a real sense already been dedicated to oil. Most of the present oil production and refinery capacity in the U.S. is located in that region. The economy of Texas and Louisiana and adjacent states is attuned to oil. Of course, exploration and development of the Gulf of Mexico must be subject to environmental safeguards and leasing policies that will assure minimum ecological damage. The record suggests that with such safeguards expanding drilling could safely take place.

Land-use policy is a major area of conflict between energy

companies and the environmentalists. Some of the nation's best remaining recreational land is being lost to energy developers. The government lacks any coherent process for deciding questions of land use. Decisions on where large electric power plants and connecting transmission lines are to be located are currently made in a jungle of local, state and federal government regulations. Legislation is urgently needed to establish an open planning process which provides for early citizen participation in deciding land-use policies. The government entity assigned the responsibility for making such decisions should also have the authority to enforce them.

The remaining open spaces in the U.S. can no longer be considered automatically available for industrial development to suit the convenience of the utilities and the oil companies. The prime missing link in the government's environmental programs is to develop a decision-making process by which the public can decide what areas are to be preserved and where development is suitable. Identifying the proper level of government for deciding these siting issues is at the heart of the problem. For siting electric power plants and transmission lines it would seem desirable to create regional siting authorities to coincide with the boundaries of the power pools in which the utilities participate. This would require the utilities to combine and coordinate their planning of new facilities. The same siting authorities could exercise similar oversight for oil refineries, liquified natural gas plants and plants that convert coal and shale into petroleum.

From an environmental perspective the evidence points to oil and natural gas found on dry land as the least damaging source of fuel. We should encourage secondary and tertiary recovery of oil from existing wells, which (as we saw in Chapter 5) could double the nation's oil reserves over the next ten years. We should also encourage a more intensive search for domestic oil and gas on shore in the U.S. This search should include the heavy oils that have been neglected and the natural gas in tight formations in the Rockies which may now be economically recoverable using conventional rather than nuclear explosions. New tax and regulatory policies to encourage such exploration and development will be discussed later in this chapter.

What about nuclear power? Should there be a moratorium on construction or should work proceed at a faster pace? On the record of the nuclear power plants operating in 1973, the nation is entitled to be "nervously optimistic" about nuclear power. But the major problems (discussed in Chapter 4) remain—power plant safety, waste disposal and safeguarding fuel against diversion. Nuclear power may well pose a lesser threat to man than air pollution from fossil fuels. We do not know. For decades the same government officials in the Atomic Energy Commission who promoted the development of nuclear energy also decided whether it was safe. The nation is moving ahead with construction of nuclear power plants on the basis of assurances by the Atomic Energy Commission that this is a safe course of action. The AEC is probably correct, but the uncertainties are far greater than the bland and usually unqualified official assurances would make it appear.

Because of these uncertainties, there is a growing belief that despite the energy crisis we should make haste slowly in expanding atomic power production. A moratorium on new plants is advocated. It is argued that the moratorium should remain in effect until the back-up safety systems have been tested, satisfactory waste disposal plans implemented, and a secure system for safeguarding nuclear material placed in operation. A further condition is that the private insurance industry should be willing to ensure the safety of the plants. Proponents insist that only a moratorium will exert sufficient leverage on the atomic energy industry and the AEC to resolve the basic unanswered questions.

The contrary view comes from those more concerned over energy supply who honestly believe that the atomic plants are safe and that remaining problems will be resolved in time. They see atomic power as a growing domestic energy source that avoids foreign policy problems linked with oil as the nation's energy workhorse after the hydrocarbon age ends.

It would be a terrible mistake to attempt to speed up the

atomic option. At the very least we must take time to assure that the plants under construction are built as safely and reliably as possible. Any new plants not now underway should be confined to "nuclear parks" where facilities to reprocess the wastes would be located on the same site as a number of plants. This would confine the radioactivity to the site and permit more effective measures to protect the public against accidents and theft. More fundamentally, what is needed is frank and honest public discussion of the risks inherent in using atomic energy and a complete review of the U.S. atomic power program by the Congress. Until an independent Atomic Safety Board can assure the public that the basic nuclear power issues are resolved, the case for a moratorium gets stronger each day.

As for coal, our best short-term approach seems to be to mine the deep coal in the East and confine strip-mining to areas where we can demonstrate effective reclamation. Once the coal is mined, the question is: Dare we burn it? The answer is a qualified Yes—if the government forces compliance with air quality standards. Technology is available to control the sulphur oxides in burning coal, but the utilities are dragging their feet in buying and using it. It is expensive; chemical engineers are needed to operate it properly, talent that power companies usually don't employ. Nevertheless, if the word were given from Washington that pollution-control technology must be used, a start could be made in installing the stack gas cleaning equipment prior to the 1975 clean-air deadline. Even if the utilities did not finish the job on time, they would have little difficulty obtaining permission to operate while they did so.

The optimum sources of domestic energy supply over the next ten years thus include oil and natural gas from onshore, the Gulf of Mexico, and Alaska; expanded coal production from deep mines and level land in the East and the Midwest; nuclear energy (if there are satisfactory answers to the unsolved problems) and whatever help we can obtain from such sources as geothermal and solar energy (see Chapter 12) and solid and organic wastes. Of course, these domestic sources will have to be supplemented by imports. Shale oil could begin to make a contribution toward the end of the 1970's, and so could synthetic petroleum made from coal. But in view of the technical and environmental difficulties these energy sources face, we dare not count on them for much assistance in the next ten years.

This raises the question of what policies we should pursue to expand the most desirable domestic supplies of cleaner energy.

Policies to Expand Energy Supplies

A national energy policy cannot depend entirely on conservation. Regardless of how much waste we can eliminate, there will still be a need for a sizable increase in the supply of energy. In the next decade, the U.S. population will increase by some 20 million, which will increase our needs for energy, housing and transportation. Many Americans need better housing and aspire to a more comfortable lifestyle, and this will require more energy. To be sure, energy conservation can offset a good part of these new needs, but over the next ten years the U.S. probably will need to increase energy consumption by at least 20 percent. This growth rate would be much slower than in the recent past, but it will require increased fuel supplies. And we must remember that as existing oil and gas reservoirs become exhausted, many new discoveries will be needed merely to maintain current levels of oil and gas production.

IMPORTS AND FOREIGN POLICY

Before the 1973-74 Arab oil embargo, the U.S. had planned to obtain most of its growing energy supply from Middle East nations. That prospect now seems like a trap, and a very costly one at that. Arab oil must be considered the most expensively priced interruptible supply on earth. If the U.S. has learned any lesson from the embargo experience it is that imports from the Arab nations must be small enough and our oil stockpile large enough so that we need not worry about another cut-off.

Imports from non-Arab nations such as Iran, Venezuela,

Canada, Nigeria and Indonesia have proven secure but are also becoming extremely expensive. Even more to the point, Canada and Venezuela are incapable of major expansion of their productive capacity in the next 10 years (except for the heavier oils) and Western Europe and Japan will be clamoring for any increased oil Iran and the other nations make available.

As a matter of policy the U.S. should confine imports of Arab oil and natural gas to a level that can be backed up by stockpiles and strategic reserves with productive capacity to equal the volume of energy imports from Arab nations. This would mean a combination of storing quantities of crude oil in tanks and preparing oil fields (such as the Naval Petroleum Reserves and the Atlantic and Pacific offshore reserves discussed earlier) for rapid production on short notice. The U.S. should move immediately to make the investment in strategic reserves necessary to match the volume of Arab oil imported. In fact oil imports should be earmarked for building up a 60-90 day stock before we go back to driving as usual. A tariff should be placed on imported oil from the eastern hemisphere to finance and maintain the necessary reserve capacity.

For the longer term the most important step the U.S. could take in its approach to the Middle East would be the measures discussed in Chapter 6 to develop an all-out program of technological assistance and encouragement of energy-intensive industry to be located in the producing nations. Providing the skilled manpower, technology and markets to provide the oil-producing nations with incentives for investing large sums of money to improve their economies is probably the best approach to avoiding worldwide shortages. It may be also the surest road to eventual peace in that troubled area. At the same time the U.S. government should take the initiative to assure early development of the large heavy oil resources in the western hemisphere. In Venezuela these projects require governmental initiative because that country is determined to develop its resources through its own national oil company. U.S. financial and technical assistance and ready access to our market could offer the

Venezuelans a very attractive package of incentives to increase production significantly within the next ten years.

Improvements are badly needed in U.S.-Canadian relations. The U.S. pursued a go-it-alone policy in planning the Alaskan pipeline and for that reason and many others now finds itself at odds with a friendly neighbor that could be a major source of fuels in the future. Having spurned Canada's requests in past years that we remove import quotas on oil, we now find Canada imposing export quotas to prevent the U.S. from getting too much of Canada's petroleum supply. And on the single energy commodity that Canada still wants to sell at low prices—uranium ore—the U.S. stubbornly maintains an absolute ban in order to protect the domestic industry.

One way to break the ice, obviously, would be to remove the unnecessary ban on importing uranium, at least from the western hemisphere. This would be more than a friendly gesture; it would get us back working together in the interests of both nations. The U.S. government should refrain from any talk of "common energy policy" or similar domineering phrases that send chills up the Canadians' backs. We should explore with the Canadian federal and provincial governments those projects—development of petroleum in the northern frontier, the tar sands, undeveloped hydroelectric sites including the large tidal power at the Bay of Fundy—on which Americans and Canadians could cooperate for mutual benefit.

We could recognize that it will be necessary for Americans to be sensitive to Canadian fears of U.S. dominance, to their hopes to expand other sectors of their economy apart from raw materials, and to their desire to have access to U.S. markets without feeling that the dollar's worth of oil they sell us means giving up the chance to sell us a dollar's worth of finished goods.

The U.S. can take several other important actions to enhance its ability to import energy from friendly nations. One is to streamline government approvals for importing natural gas. Hearings before the Federal Power Commission now take years, which means prospective U.S. purchasers lose out in most of their attempts to buy gas abroad. The idea that the FPC can influence what other nations charge for their gas is somewhat ludicrous. Natural gas imports should be under the same White House ground rules as oil. Approval should come from the State Department with the FPC in the role of technical advisor concerning the need for the gas at the negotiated price. Questions concerning the environmental impact of new LNG (liquefied natural gas) facilities would be better handled by EPA working with the affected state. There should be advance planning and of course a public hearing, but a decision should follow promptly by the EPA Administrator.

The crucial issue arising from future LNG imports is the same as with oil—the foreign policy benefits or dangers flowing from the transaction. In some cases the government may wish to approve a transaction even though the FPC believes the price is too high. For example, high-priced imports of gas from the Soviet Union may offer overriding benefits in non-energy areas that justify approval. Certainly the decision whether to turn down such a transaction, which may be part of an overall settlement of the cold war, should be made by the President, and not determined by the Federal Power Commission.

A significant portion of U.S. oil supply—perhaps some 30 percent—will continue to be imported in the next ten years. But certainly the bulk of our energy and a large fraction of our growth must come from domestic sources. It is to our economic and foreign policy advantage to explore energy options with this priority clearly in mind. Minimizing the U.S. drain on the world energy supply may be the most constructive step we can take to help Japan, Western Europe and the Third World. If we import less there will be more for them and perhaps at lower prices. International cooperation to help other nations develop whatever energy resources they may possess will also help each nation balance its energy budget.

The energy problem is a global issue. It can be a means of loosening some of the narrow reins of nationalism, of bringing the people on earth closer together in peaceful pursuit of a common concern. Or, as we have seen, it can divide and torment us.

Oil and Gas

Most people are surprised to learn that the greater part of the oil and gas in the United States that is economically attractive to develop in the next decade is actually owned by the American public. The most effective action the federal government could take to enlarge the supply of natural gas and crude oil would be to launch an intensive effort to measure reserves, assess the environmental problems and then drastically to reshape its leasing program accordingly. At the moment efforts to speed up the leasing program risk a give-away of the people's resources and destruction of fishing and recreation values because the federal government employs such a small force to appraise and lease the federal land. Armed with accurate information, and a public interest evaluation of the conflicting claims on the land, the resource could safely be brought to market under an expanded leasing program much more rapidly.

Reforming the leasing program is the key to expanding offshore production and developing strategic reserves. We need a leasing system that stresses information-gathering by the government, production by the oil companies and profit-sharing by the government. Under the present system the government is not sure of the pace or extent of development. And if prices go up as they have recently, the oil companies pocket 83.5 percent of the money and the federal government only 16.5 percent. Only in America do the people sell their oil in the ground at a fixed price and buy it back at skyrocketing prices.

A second major policy reform would be for the government to provide new incentives for fuller exploitation of the oil and natural gas resources on dry land. The incentives would apply to any natural or synthetic petroleum transported to U.S. markets via an all-pipeline route. North American land-base petroleum development would eliminate the dangers of an ocean voyage for that segment of our energy supply and thus alleviate the growing pollution of the oceans from oil spills. Even more important, it would relieve the pressure for industrialization of the coastal zones. Refineries could be economically located inland with much more flexibility in selecting sites and without the intense pressure for coastal zone development which is inherent in offshore drilling or imports.

A North American land-based petroleum policy would offer obvious foreign policy advantages as well by eliminating ocean transportation. It would give us a most secure source of oil in wartime. And, of course, to the extent such resources can be economically developed it would minimize concern over our inability to import more oil from overseas. There is little debate on this score. The only real question is whether such an option holds substantial promise for the next ten years. No one can be certain of the answer, but most of the remaining natural gas in the U.S. is beneath dry land. The reverse is true for oil, but a stepped-up drilling and recovery effort could at least keep land-based oil production stable for the next decade rather than allowing it to continue to decline.

There are three major areas of federal policy that should be reshaped to give land-based petroleum production a boost —the tax laws, price controls and the strengthening of competition:

- Tax incentives for petroleum should be confined to land-based domestic production with larger incentives for secondary and tertiary recovery and for production of the heavier oils and natural gas in tight formations.
- Price controls should be removed for all new production of such oil and natural gas.
- Tax incentives and freedom from price control should be made available only to companies that agree to confine their business to the land-based production of U.S. petroleum and agree to reinvest all of their earnings except reasonable dividends in such production.

The tax laws should be changed to confine the special provisions for oil and gas to production above the high-water mark in the United States and in adjacent nations where the petroleum is transported to the United States via a land-based pipeline. Legislation should also revise the application of the depletion allowance and intangible drilling expenses to provide an equally effective tax break for synthetic petroleum made from tar sands, shale oil or coal (as well as extending similar tax treatment to competing forms of domestic fuel such as uranium or geothermal steam, and new forms such as solar energy or fusion). This measure is extremely important because the present law actually handicaps new forms of energy as compared to conventional sources of petroleum. The legislation should also contain special new incentives for secondary and tertiary recovery and production of heavier oils and natural gas in tight formations that otherwise might not be economical. The additional incentive would equal the additional cost of such production, within reasonable limits, and thus make recovery economical except for very marginal resources.

New reform legislation (for the reasons discussed in Chapter 8) should eliminate the special tax provisions applicable to overseas production, including the tax credit for payments to foreign governments (except Canada and Mexico), thus greatly strengthening the incentives for drilling in the United States. By confining tax incentives to land-based areas such drilling would become much more attractive. If the payment of additional taxes causes increases in the price of crude oil that will also serve to help encourage energy conservation. By putting all energy sources on the same footing, these tax reforms would enhance the economics of shale, geothermal energy and gas or oil from coal. In addition, special incentives keyed to the marginal costs of secondary and tertiary recovery would make it feasible to obtain a great deal of additional production from existing oil reservoirs, a good source from an environmental standpoint.

These tax incentives should also be confined to companies that are now independent producers of oil and natural gas only in North America or companies that might be formed to undertake such activities. To be eligible for the tax incentives in the future, companies would need a charter specifically limiting their activities to land-based petroleum production in North America with a requirement that all of their earnings, except reasonable dividends, be reinvested in such petroleum production.

As a result, the integrated major companies would no longer be able to use any tax advantage from their production operations to give them a break over independents in the refining and subsequent retailing of oil products. All of the benefits of the depletion allowance and intangible drilling expenses would be confined to encouraging more production, and they would not be available to companies that wished to engage in refining and other down-stream operations. Refiners and distributors would then be on an equal footing so far as the tax laws were concerned. This would be a giant step to strengthen competition in the petroleum industry.

The regulatory part of the package of incentives would exempt from price control all new sales of natural gas (or oil) by independent producers in the same land-based areas. The natural gas and crude oil now flowing to consumers would be placed under price controls that fixed the price of flowing gas and oil at levels related to costs, a consumer protection measure to be discussed later in this chapter. Removing controls from the price of new discoveries would give a boost to deep-drilling for natural gas in the Southwest and encourage the development of natural gas and oil resources in Alaska as well as the large gas resource in the tight formations in the Rocky Mountain area.

There is every reason to continue price controls on natural gas in the federal domain off shore. As we concluded in Chapter 7, the consumer cannot rely on price competition among the producers to assure reasonable prices for the foreseeable future.

The major integrated companies dominate gas and oil production in the outer continental shelf. These offshore operations are undertaken in quest of oil, and the interest of the majors is primarily determined by the price of oil, not natural gas. This is likely to continue to be the case because experts in the U.S. Geological Survey believe 90 percent of the remaining offshore natural gas is associated with oil.[257]

The evidence refutes the popular myth that the Federal Power Commission's price controls on natural gas have inhibited the search for natural gas in the federal domain off shore. Every time leases have been offered by the federal government, the major oil companies have bid billions of dollars for the privilege of seeking oil and natural gas. Since 1954, for example, companies have paid bonuses for leases to develop oil and gas on the outer continental shelf of almost $10 billion. The fact that oil companies and others are willing to bid such huge sums to look for gas that must be sold under FPC price controls belies the claim that price controls are shutting off exploration. If there is any lag in the development of these offshore leases—which the companies deny—then the remedy lies in stronger management of the government's leases as previously discussed. The government can spur development by issuing leases that require an all-out production effort.

The nation does not need to bribe the oil industry to develop the resources owned by the public. The only limiting factor is the pace at which the Department of Interior can learn enough about the Gulf Coast to offer more leases, under terms which will ensure that the oil industry does not simply sit on them. And as previously discussed, the offshore Atlantic and Pacific should be developed as strategic petroleum reserves for future use.

Why, then, should price controls be removed from the sale of newly-discovered gas by independent producers on land? The basic reason is that higher prices on shore will result in the remaining natural gas being produced more quickly. On land most of the gas is separate from oil and the rate of drilling would be responsive to a higher free market price. Moreover, the onshore gas can be sold intrastate, unlike gas from the offshore federal waters which must cross state lines and become subject to FPC price controls. If price controls are not removed from new supplies of natural gas on land, this gas increasingly will be sold intrastate at free market prices.

Natural gas production on land, unlike costly offshore operations, is not a business for Big Oil only. On the contrary, most of the land-based explorers are independents. Entry into the business is relatively easy. Decontrol would encourage a stronger drilling effort on land, unlike the outer continental shelf where available acreage from the government is the limit. Decontrol also would provide greater economic incentives for developing the gas in tight formations in the Rocky Mountain area as well as enhancing the prospects of synthetic gas. Legislative decontrol of prices for sales of new gas supplies by independent producers holds out the prospect of stimulating competition and eventually bringing prices into line with costs.

There remains the problem of the windfall profits that would accrue to the exempted independent natural gas producers. The requirement that free market prices will be available only to companies confined to land-based U.S. oil and gas production insures that these profits will go back in the ground to produce an enlarged supply of secure energy. This is another reason why the major integrated companies are excluded from the exemption, for they could use the profits to buy service stations, real estate or to drill in Saudi Arabia.

In time, competition may eliminate windfall profits. If not, at least the consumer will share in the benefits under this approach. The alternative of the federal government entering the oil business throughout the country is easy to suggest. Realistically it is an answer likely to produce debate, but little oil.

For several years petroleum production has been declining. Yet enormous quantities of oil and natural gas remain under American soil. As we have seen, there are limits to the pace at which these resources can be produced in a socially acceptable way. But a combination of selective acceleration of offshore and onshore development can at least halt the decline and perhaps bring a modest increase in production. Domestic oil and gas can fill part of the gap that otherwise would be filled by Arab oil in the next ten years. And coal can supply whatever else is needed to meet the needs of a conservation-oriented pattern of U.S. energy growth.

Coal

A plan for the use of coal is a key part of a balanced energy budget that will enable the U.S. to pursue an independent foreign policy in the next ten years. And coal can serve the interests of environmental and consumer protection if used in a sensible and selective way.

There are two environmentally satisfactory sources of enlarged coal production. One approach is to achieve increased production from existing deep- and strip-mine operations. To do so we must solve the problems which are creating disruptive labor unrest. Wildcat strikes and absenteeism result in an effective four-and-a-half day work week in the coal industry. Unless management adopts a more safety-conscious and profit-sharing attitude, we face a deepening crisis in coal production. With the price coal now commands, the mines can be made safer and labor can be paid more generous salaries and fringe benefits. Given labor peace in the industry, coal production can be increased 20 percent over 1973's output without opening new mines.

To fill the energy gap for the next ten years, new mines will be needed. The strippable coal resources in the Illinois Basin are the most promising large scale source of secure energy supply available for quick exploitation. Such mining can begin operations in two or three years if the manufacture of the big shovel receives priority treatment. We can also open new deep coal mines in the East, but these require four years or more to go into production, and labor will remain scarce until we learn to mine coal more safely.

There are billions of tons of coal in Southern Illinois, Ohio, Western Kentucky and surrounding areas, and production could be enlarged by tens of millions of tons each year. We know that it is possible to reclaim these fairly level lands in an area with good rainfall if stern laws require it and if the reclamation effort is sustained over many years. This coal is reasonably close to industrial centers and is now much lower in cost than imported oil.

Expanded coal production in the Illinois Basin would be used to convert to coal most electric power plants and large industrial establishments that now burn natural gas and oil simply as sources of heat. The nation currently uses more than 25 percent of its total energy in the form of natural gas and oil as sources of industrial heat which could be replaced.

If such a replacement program were implemented over the next ten years, large quantities of oil and gas would be available to supply transportation and small residential and commercial needs. The coal in the Illinois Basin is high in sulphur content and therefore an integral part of this program would be the installation of equipment to prevent the ash and sulphur in the coal from escaping into the air. Such equipment has been developed and it needs to be put into commercial use.

A national commitment of the highest priority should be made to accomplish this switch to coal over the next ten years. It is one of the few feasible options available to the U.S. for becoming fairly independent in our energy supply without sacrificing other values. It does not rely on new discoveries in the ground or breakthroughs by the research establishment, which may not occur. In the next ten years burning coal as coal is a dependable option that will require a special effort—priorities for shovels, coal cars, pollution controls, reclamation and mine safety. But it can be done.

Atomic Energy

For reasons set forth earlier, a crash program to accelerate atomic power is too risky an option. Nevertheless, the completion of power plants under construction and planned new starts, built with painstaking attention to their safety, will add significantly to our energy supply over the next ten years. If atomic power proves to be safe, it can supply about ten percent of total U.S. energy needs a decade hence. To put this figure in perspective: Atomic power could provide as much as half of the new sources of energy supply needed if growth is limited by a conservation policy.

Summary

Without attempting to project the future, a balanced energy budget for the nation over the next ten years could be expressed this way:

IN QUADRILLION BTU'S

	1973	1984
Demand	75	90
Supply		
Domestic Oil	22.0	22 (11 million barrels/day)
Imported Oil	12.5	10 (5 million barrels/day)
Natural Gas	23.0	22 (22 trillion cubic feet/yr)
Coal	13.5	22 (900 million tons/yr)
Hydro	3.0	3 (56,000 mw of capacity)
Nuclear	1.0	8 (152,000 mw of capacity)
Solar, organic and solid wastes, geothermal and shale	—	3
TOTAL	75.0	90

RESEARCH AND DEVELOPMENT FOR NEW SUPPLY

In Chapters 11 and 12 and indeed throughout this book the possibilities for new energy supply technologies have been described. The failure in the past decade to push ahead on these new sources, except for nuclear energy, is at the heart of the problems we now face. There are few quick fixes and many of the basic new sources may take decades to perfect. Even so, a significant number of promising technical options could be perfected within the next 10-15 years. Large quantities of commercial energy would be forthcoming in 1985 and beyond if the nation committed the funds and provided strong leadership to do the job. By 1984 we could develop:

- New automated underground coal mining systems that would be relatively safe and economical as well.
- Commercial plants for gasification of coal to supplement natural gas as a fuel.

- Commercial plants for converting coal to oil.
- A commercial shale oil industry producing 1 million barrels per day.
- Commercial geothermal energy providing significant quantities of electric power production in the West and perhaps elsewhere.
- Commercial technology to make oil or gas from organic wastes and to use solid wastes for electric power production.
- Commercial operation of low btu gasification of coal in large-scale gas turbine-steam turbine combined cycles for more efficient electric power generation with a minimum of pollution.
- A fluidized-bed boiler to burn coal in electric power plants and industry more economically and without emitting sulphur and other pollutants into the air.
- A variety of effective techniques for controlling emissions of sulphur when burning high-sulphur coal at electric power plants.
- Dry cooling towers to permit power plants to be located at remote interior sites without using water for cooling.
- Super-conducting cryogenic electric power transmission lines to cut down on losses and permit underground construction of power grids.
- The commercial use of solar energy for direct heating and cooling of homes in many parts of the nation.
- An electric automobile for suburban use in conjunction with mass transit to use electric energy for transportation and thus reduce the need for importing oil.
- A commercial fuel cell for total energy installations that would be more efficient than existing means of energy conversion.

There are many other opportunities—the list is by no means exhaustive.

The more fundamental options for the future—central station solar energy, fusion power, nuclear breeder reactors, and low-cost means of converting energy to hydrogen—cannot be completed in the next decade. Yet they must be pursued with an all-out commitment *now* if they are to be available to ease the energy shortage throughout the world in the decades ahead. Unfortunately, with the possible exception of the nuclear breeder, none of these options is being pursued with a sufficient sense of urgency.

There is no longer a debate over whether the nation needs a stronger energy research and development program. This is one of the few points on which the energy industry, the government, environmentalists and concerned citizens all agree. Yet the flow of new technology to produce energy in an environmentally acceptable way is still running like a dry creek. In the past the problem was that the work was scattered in small offices in large agencies such as the Interior Department, the Environmental Protection Agency and the National Science Foundation. Nuclear energy has been developed only because a single federal government was assigned the leadership role as its priority mission and given the money and support from Congress to push a strong program in cooperation with industry. There was no such leadership for the other energy sources in the past decade and very little progress has been made. In fact, one would be hard pressed to name any significant commercially-available energy technology that has resulted from the other scattered R & D efforts in recent years.

The nation must make a commitment to developing solar energy, a commercial fuel cell, geothermal power and many other promising ideas. A first step would be the creation of an Energy Research and Development Agency. The President has pledged a $10 billion effort over the next five years and Congress is considering a $20 billion effort. But unless a new agency brings fresh talent with a sense of commitment to the job, this money could go down the drain. Without new leadership the existing sources of energy will consume the money. The new agency must be the champion of the clean energy options that have no commercial sponsor—the sun, geothermal energy and new technology to use energy more efficiently.

The present situation of a scarcity of fuels is due in large part to our neglect of technology. Industry and government alike have failed to mount the kind of effort needed to make clean energy available in adequate quantities. Industry's neglect is understandable because the energy-producing industry is not being hurt by the scarcity of fuels. Short supplies and higher prices suit the producers fine.

The public has an enormous stake in quickly perfecting new technology to transform abundant resources in the ground and from the sun into abundant supplies available in the market place. But perfecting technology for an adequate supply of clean energy sharply conflicts with the energy industry's incentive to live with the current technology and profitable continued scarcity. The nation cannot expect the oil companies to perfect solar energy or other sources to compete with their own petroleum reserves. There could be no more rewarding investment for the consuming public in the next ten years than to invest 1 or 2 percent of their energy bills in a strong and urgent research and development effort.

SUPPLIER OF LAST RESORT

As the nation enters the era of energy scarcity, a gaping hole has appeared in our national energy policy. No one is finally responsible for delivering the goods. Electric utilities are required by law to make the investments necessary to keep the lights on. But no one has any legal obligation to sell gasoline or heating oil, or to produce any of the basic fuels. Oil companies and coal producers place profits for their stockholders ahead of supplying their customers. They drill for oil or dig for coal only if they think the price is right. Under existing law, energy producers may sit on their reserves if they believe they can make more money developing them later. Oil companies can fail to build needed refineries, and the public can only do without the gasoline and fuel oil it needs.

Ordinarily the public could expect competitive market forces to assure that supplies offered for sale would be ample. But bitter experience has shown us that clean energy supplies are scarce in relation to potential demand. As discussed earlier, government has become less able to carry out programs or enforce laws that Big Oil disapproves. We have already seen how programs to protect the environment are giving way to fear of shortages. Within the past year, top officials in government have relaxed price controls for the same reason.

The government's knuckling under to Big Oil is typified by an episode in 1973 when Deputy Treasury Secretary William E. Simon—who soon became the President's "energy czar"—urged the Chairman of the Federal Trade Commission to back off on an anti-trust suit that the independent FTC had filed against eight major oil companies. Simon said it "gives me a great deal of concern because of its implications for domestic energy supplies." The FTC complaint asserted that there is an anti-competitive market structure in oil refining. Simon stated that the oil companies were "now having second thoughts" about building new refineries as a result of the FTC complaint. He thus warned that if the FTC persisted in its duty to enforce the anti-trust laws, the energy crisis might be worsened. Government policy on environmental protection, price controls, foreign policy and law enforcement simply cannot be dictated by fears of the consequences on fuel supply.

The U.S. government must become the fuel supplier of last resort. To the extent that private companies fall short of delivering fuels or building refining capacity, the U.S. government clearly must develop the capacity to take up the slack. The government would not compete with private companies but would only fill the gap where they failed. Simply forming a federal corporation with such power might well mean that it would never have to build a refinery or produce a barrel of oil. But its establishment would play a vital role in eliminating the growing popular fear of shortages and disproving the seeming impotence of government to act in the public interest.

The U.S. Fuels Supply Corporation would be enpowered to develop detailed projections of energy demand and to obtain the detailed commitments of the industry concerning its programs for new production and construction of refineries.

The Corporation would then be empowered to take whatever actions were required to assure that the nation's energy supply was adequate with minimum damage to the environment. It could buy energy from abroad, contract for development of fuels on federal lands (such as Naval Petroleum Reserve No. 4 in Alaska), build refineries, develop synthetic plants or do whatever seemed necessary to meet the nation's needs. It would be charged with developing a strategic oil reserve in the Atlantic and Pacific off shore.

The U.S. Fuel Corporation would operate pursuant to these basic criteria:

- It would not undertake any activity unless it was necessary to avoid a clear danger of shortages of fuel at reasonable prices that private enterprise had failed to remedy.
- It would be responsible for assuring the development of the stockpiles and strategic oil reserves previously discussed. It could explore the federal domain to better ascertain the fuel resources in the ground the government already owns and to develop such resources to the extent private enterprise does not do so.
- In any exploration and development contracts and sales of facilities or fuel, the Corporation shall give preference to independent companies that are not affiliated with major integrated petroleum companies.
- The Corporation shall be provided with initial capital by the federal treasury, shall be authorized to issue bonds, and shall be required to sell its products at prices that covered all of its costs as well as allowances for the value of federal mineral rights it acquired and the taxes a private company would be obligated to pay.
- The Environmental Protection Agency in consultation with appropriate state governments shall have the power to review and approve each action of the Corporation prior to its initiation to assure that no important environmental values would be seriously impaired.

Creation of the U.S. Fuel Supply Corporation would not signal the demise of private enterprise in the energy business. On the contrary it could be the single action that would save the U.S. energy industry from nationalization a few years hence. It would serve the double purpose of fixing responsibility in government to avoid shortages while giving industry every incentive and opportunity to do the job itself.

CONSUMER PROTECTION

The basic policies that will best serve the consumer in the next ten years are also those which will enlarge the supply of clean energy. Until supply is increased the consumer will pay through the nose. Thus the R & D effort, the various other actions to enlarge supply, and especially the establishment of a U.S. Fuel Supply Corporation will tend to reduce the enormous bargaining power that producers throughout the world now enjoy.

Urgently needed is a new charter for the Federal Power Commission, directing that agency to regulate on a strict cost basis the natural gas and oil already discovered and flowing to the consumer. The issue is clear-cut and it poses a multi-billion dollar question: Will the petroleum industry be allowed to charge higher prices on oil and natural gas whose costs are virtually constant, simply because these higher prices can be exacted in a sellers' market? The fact that higher prices may be justified to cover the costs and provide incentives for new discoveries is no excuse for price increases on production begun years ago, which is profitable without multi-billion dollar increases in prices.

The potential windfalls are enormous. The immediate windfall in petroleum is some $5 billion a year as prices for already discovered oil climbed from about $3.40 a barrel in 1972 to the $5.25-a-barrel price the Nixon Administration has sanctioned. As for natural gas, each price increase of $.05 per thousand cubic feet adds another billion dollars to the consumers' bills. We have previously outlined a package of incentives including removal of price controls to spur new discoveries by independent producers. But consumers need protection against the market power of producers to change today's high prices for oil and gas discovered years ago. Unless these prices are rolled back and held in line, the petroleum companies will continue to reap many billions in windfall profits each year.

The word windfall is used advisedly because there is no other way to describe the transfers of money from consumers to the petroleum industry that have no cost basis but reflect only the sellers' bargaining power in a market where prices are skyrocketing. This kind of fuel pricing only feeds the flames of inflation. The oil and natural gas involved has already been discovered and it is now flowing to the consumer. The companies are not obligated to put the money back in the ground, and their incentives for doing so are determined by the price of new supplies they would be searching for, not the price of petroleum already discovered.

Price controls on these basic fuels will not help the consumer much unless there are effective controls on the retail prices the consumer pays. Price controls will be needed on gasoline and heating oil in the coming years of shortage. Otherwise, the prices of these essential items could soar out of sight. Price increases should be limited to increased costs —that is, to *real* costs actually incurred.

It also is important to strengthen the existing regulatory system of price controls on the retail price of electricity and natural gas. These are now administered by state utility commissions, but there is increasing concern over whether the states can handle the job, especially the fixing of rates for electric companies tied into interstate power pools. The thrust of the regulatory effort should be to ensure that the utilities scale their expansion plans down to what is actually needed under a national policy of conversing energy. This is very important to consumers because rapid expansion simply means higher rate increases. As we saw earlier, for the next ten years new sources of fuel and new power plants will be more expensive than existing sources. Thus a major element of consumer protection would be for regulatory agencies to require the utilities to "promote" conservation as ardently as they stimulated growth in the past. This will be no small task.

The time is right for a major Congressional study of the structure of the electric power industry, to determine whether new power plants and transmission grids are being built at maximum efficiency.

The present industry structure is clearly inadequate. Region-wide generation and transmission companies, privately owned but with some public ownership and participation in the management, would be a tremendous improvement. These companies would be a natural outgrowth of the present power pool arrangements where coordination takes place through coordinating committees. The difference is that a full time corporate management representing all the power companies in the region and the public at large would have the responsibility and the authority to plan, build and operate a region-wide grid. Sites would be selected on the basis of the best land-use planning for the region as a whole. Duplication would be eliminated and reliability of service improved.

What is suggested here is but the bare outline of a system in which "Comsat"-type generation and transmission companies would sell electricity at wholesale to the existing utilities as distributors. This outline requires detailed study to chart a definite course of action. A congressional mandate for the Executive Branch to make such a study and implement the results is clearly needed. Such a study could explore, among other things, the concept of regional companies, with some government participation, that would be responsible for generation and transmission lines while existing companies remained in the retail distribution business. Quite apart from a congressional study, federal legislation should be enacted to authorize regional siting and rate-making boards; to require the utilities to engage in long-range, well-publicized planning; and to require that all new sites and facilities be approved by the siting board before construction could begin. The siting boards should be given comparable oversight for oil refineries, synthetic fuel plants and other energy developments.

These same regional power plant siting authorities could also play a role in reforming the rate-making process for the electric power industry. In order to perform the siting task the board and its staff will need to become intimately familiar with the bulk power supply systems of the region and to consider the costs of various alternatives. The staff resources necessary for performing the siting task adequately could thus also help determine the appropriate cost of bulk power service and the rates utilities may charge. The regional siting boards should be further empowered to lay down guidelines for retail rate-making that incorporate the energy-saving concepts discussed earlier as part of a policy of conservation. They could also be authorized to determine the regional cost of bulk power service for rate-making purposes. The state commissions would continue to review the actual distribution costs within their jurisdiction and prescribe rates pursuant to the guidelines established by the regional agencies.

There are also actions that can strengthen competition in the fuel industry and these are of prime significance to consumers. The FTC suit, mentioned earlier, will eventually resolve the basic question of whether the large integrated oil companies should be broken up. But consumers need action now to make the present fuel industry more competitive. A first step would be to enact the tax reforms that put all producers, independent refineries and new companies on the same competitive footing. Other reforms to enhance competition are possible in federal leasing policy. Oil and gas leases now are sold to the highest cash bidder, which usually means a large integrated oil company or several of them operating in a joint venture. The alternative of royalty bidding—where the company offering the highest royalty to the government receives the lease—should be tried, perhaps combined with guarantees concerning rapid development. In that way smaller companies could more easily participate in the bidding.

Last but most fundamental to consumer protection is the need for government to get the facts about energy and make them available to the public. Nothing erodes public confidence more than the lack of reliable information about how much oil is available, how much it costs, the size of oil and gas reserves, and a hundred other basic facts about the industry. In the winter of 1973-74, the government was unable to determine the size of the energy shortage, which cast grave doubt on the government's ability to cope with it.

To the extent the government now has information, it comes almost exclusively from the energy companies and trade associations. Obviously statistics on how much oil a well is pumping must originate with an oil company. But government must develop the capability for independent gathering, analysis and evaluation of the basic facts on which to make energy policy. The energy companies must learn to operate with their books open just as the electric utilities have done since the reform of the utility industry in the 1930's. The establishment of a Bureau of Energy Statistics with the competence and integrity of the Bureau of Labor Statistics is a vital part of a national energy policy capable of serving the public interest.

CHAPTER 14

1985 and Beyond

The year 1985 sounds quite distant, but "the future is now" because of the long lead time we need to develop new, cleaner sources of energy and change our consumption patterns to achieve large and significant savings. If we are to balance U.S. energy supply and demand in the decades ahead in an environmentally acceptable way, we must make the basic policy choices and decisions now. We face the imperative that Edmund Burke stated well almost two centuries ago when he wrote:

> The public interest requires doing today those things that men of intelligence and good will would wish, five or ten years hence, had been done.

Events in the winter of 1973-74 brought the United States to a crisis, or turning point, in its pattern of energy consumption. The decisions we make in the mid-1970s will largely determine our national energy policy in 1985 and the years beyond. These decisions in turn will go far toward determining the kind of cities and countryside in which Americans will live, and the quality of our lives.

I believe a solution to the energy related problems will require profound changes in the material growth ethic that now prevails in this country. Different individuals will, of course, see our national situation differently and propose different courses of action. This chapter briefly sets forth my personal assessment.

One alternative, obviously, is to attempt to continue our consumption habits and patterns formed in the past era of abundance—a "back to normal" approach. To satisfy such demands, an all-out effort to develop domestic energy sources must be mounted at once. This course rejects as defeatist a policy of energy frugality and assumes that supply can be made sufficient to continue our present energy-intensive American economy without constraint. Such a course also subordinates the environmental and foreign policy problems inherent in sustaining our accustomed 3 to 4 percent growth in energy consumption far into the future. It further assumes that after a brief period of "belt tightening" we will somehow gain the new technology necessary to supply this growth indefinitely and to control resulting contamination of the environment.

The primary purpose of a "back to normal" approach is to preserve and expand the key features of present-day middle-class American life—unrestricted use of a large family car, comfortable enjoyment of a detached suburban home and expansion of an economy geared to producing more and more material goods. The highest priority in such a "normal" policy is to enlarge the supply regardless of direct and indirect social costs. Such a policy recognizes no limits on growth.

No one can be certain that such a future is impossible, and to many it may seem the most desirable option. "Normalcy" has great emotional and psychological appeal after a decade of crisis and upheaval. Yet the assumption that we can continue the high energy growth rates of the past beyond 1985 is actually very risky. Such a policy rests on the belief—amounting to a perilous act of faith—that the untried and difficult technical options for expanded energy supply discussed in Chapters 11 and 12 will in fact succeed and produce large volumes of energy in the 1985-2000 period, and that the related environmental concerns somehow will be satisfied.

Another alternative, at once more conservative and realistic, is to reduce our energy demand to the level of supplies that we can reasonably and safely expect to have available in 1985 and beyond. This approach of intelligent austerity will require a willingness to make rather fundamental changes in our economy and our values. For as we look past 1985, it may not be sufficient merely to squeeze the energy waste out of our present way of life using the energy conservation policies set forth in Chapter 13.

Growth in energy consumption reflects the increasing number and size of the things we own—and of the things that own us or at least claim a large part of our time and attention. So long as the United States continues to assume that "more is better," all our efforts at increased energy efficiency—small cars, mass transit, industrial re-engineering—can achieve is to buy five, ten or perhaps fifteen years of additional time. Thereafter the demand for energy will resume its former rate of exponential growth and soon confront us with the same problems we had sought to escape but merely deferred and intensified.

The energy policy issue, then, is fundamental, demanding that we reassess our definition of "growth," our criteria of individual and collective well-being and even our ideal of the American way of life. Many times in our history we have changed, adjusted and matured, emerging stronger. Now we are entering another time of change, another test of our intelligence and maturity.

My study of energy policy has led me to the most difficult and troublesome conclusions: That the United States economy must undergo fundamental restructuring in order to balance our energy budget. If we are to reach a stable level of energy consumption by 1985, the mix of the nation's GNP must undergo a major change. ("Zero-energy-growth" is a misleading phrase—it will require huge and increased quantities of energy to sustain America at high levels of consumption in 1985.) Yet a policy of stability does imply that we are fast approaching the saturation point for material goods, and that the emphasis of our economic growth must shift to the encouragement of services and other occupations which do not require as much energy as manufacturing.

The United States sets the pace and determines the shape of the energy growth curve throughout the world. Therefore, the problems posed by a U.S. energy demand that grows by even 2 percent a year after 1985 are enormous, especially since we dare not consider our needs in isolation. The global energy needs of billions of people are at issue, and their supply is very much in doubt. The cruel impact of skyrocketing oil prices on India, and the dearth of petrochemical fertilizers in Southeast Asia merely foreshadow the tragedy of global energy shortages that can be expected in the 1980's and beyond.

We are concerned not only with future shortages (although, as we saw earlier the possibilities are real) but also environmental dangers. A basic reason for changing the pattern of our economic growth to reduce energy consumption is to forestall such environmental dangers as air pollution and the ravaging of our coastlines and countryside. We do not wish to consume much of our heritage to produce more energy that yields no lasting and compensating benefit. Ultimately, we are concerned that we may appease our energy appetite by changing the earth's climate in an unpredictable way, and upsetting the ecology of the earth as a whole.

Placing a ceiling on growth may be desirable as well as necessary. Americans have voluntarily reduced birth rates so drastically that we may approach a stable population early in the next century. Since Americans consume six times the resources as the worldwide average, this trend toward zero population growth in the U.S. is of great importance. Even now a growing number of Americans are finding that happiness does not depend on more and more material goods. Of course, many Americans have yet to attain a decent standard of living, much less a life of affluence; and most people in the world are in poverty. For them, economic growth is an urgent necessity. But for many Americans who realize they have reached a saturation point in per-capita consumption, more growth does not mean more happiness and fulfillment.

Politicians, social theorists and ordinary citizens throughout the developed world are beginning to question increasingly the familiar material growth ethic. A remarkable testament to this changing perception is Japanese Prime Minister Kakuei Tanaka's best-selling book, *Building a New Japan*.[258] No country has achieved such rapid rates of economic growth as Japan in recent years, and perhaps no country has sacrificed so much of its environment and human values in the process. Tanaka's plans would mean a sharp change of pace for Japan. He projects a future industrial structure based not only on quantitative growth but also on improved quality of life in Japan. As he writes:

> . . . the industries to lead the economy in the future should be selected using the criteria of how little damage they do to the environment (pollution burden standards) and how much pride and pleasure they give their workers (labor environment standards)—the centre of gravity of the new industrial structure should be changed from material and energy consuming heavy and chemical industries to knowledge intensive industries making greater use of man's wisdom and knowledge. . . . It is these knowledge intensive industries which serve the cause of harmonious co-existence between industry and environment and are the key to recovering our humanity.[259]

In this country, too, more people are becoming critical of the headlong rush for growth and are seeking a more orderly and balanced concept of national development and individual fulfillment. In part, this attitude may spring spontaneously from the changing structure of our society. Professor Daniel Bell in *The Coming of the Post-Industrial Society*,[260] sees strong tendencies for the U.S. to move away from a traditional, private enterprise goods-producing economy, and toward a service-based economy with an expanding public sector. He anticipates that the market increasingly will be dominated by social rather than economic decision-making, a development John Maynard Keynes also anticipated.[261] The energy crisis may well accelerate this trend. The problems now facing us in the energy field form a microcosm of the problems we shall soon face in every sector of the economy and society.

The end of the era of low-priced energy in the United States and other industrialized nations could result in a worldwide redistribution of industrial growth during the next decade. Much of the growth in energy-intensive basic industry and manufacturing is beginning to take place in the energy-rich developing countries.[262] For three decades such industries have clustered along the Tennessee and Columbia River Valleys and the Texas Gulf coast because low-cost energy was abundantly available. It is only natural that this industrial growth should now shift to the Persian Gulf and other sources of abundant low-cost energy. We should be prepared for such diversification of the world economy. The developed countries may well be forced to concentrate on those technological, knowledge-intensive industries in which they have the greatest advantage. Within this global division of labor, the U.S. could be expected to concentrate on food production, for which this country has important natural advantage, as well as the high-technology and service industries in which we excel.

For these and other fundamental reasons, the growth rate in U.S. energy consumption in the period 1985-2000 is very much an open question. In my view a realistic and desirable objective for the U.S. would be a full-employment, knowledge-intensive, food- and service-oriented economy that would be fueled at a fairly stable level of energy consumption. Such a society would contain a high degree of material affluence and consume huge quantities of energy each year. But the U.S. would be able to demonstrate how much affluence is enough and give other nations an opportunity to use more energy and catch up.

To the environmental and political constraints on our growth in energy production should be added a moral constraint. If all the peoples on earth are eventually to enjoy an adequate level of material well-being, we will need to adopt a new ethic which regards waste as a form of theft. For if we continue a self-indulgent, disposable society where the cycle continually is to dig, burn, build and then discard, we are stealing from our children and grandchildren the planet's resources. Our energy appetite and its consequences should cause us to question the whole structure of our material-oriented, energy-intensive pattern of industrial production, and at bottom, the basis of our civilization.

A root fallacy in the present organization of American economic life is vividly demonstrated by the manner in which we measure productivity. Mass production is the overriding criterion. Success is determined by the number of automobiles or shoes or light bulbs produced per unit of investment in capital and labor. Yet, if we think about it, the number of light bulbs a worker can produce in a day and the total manufacturing cost of a bulb are not the decisive tests at all. What is more important is how long the bulb will last and its cost per hour of use for the consumer. We must question the system which measures productivity by the unit cost of production to the manufacturer, rather than the unit cost of service to the consumer.

Emphasis on quantity rather than quality and durability dominates our contemporary society. And as long as the supply of material goods and energy to produce them seemed infinite and the environment a free garbage can in which to dispose of them, the system made sense. But we are misled by a very incomplete accounting system. Consumers subsidize the production costs of industry because these costs do not include such items as the damage to the environment. Nor does the price tag reflect the cost to society of disposing of poorly built automobiles, toasters or tricycles after they fall apart.

Our economic system works to make it profitable for the production of new goods to increase. The system has performed quite well in achieving our current standard of living. But we can now perceive a need to conserve our resources and protect the environment. Therefore, it is time to examine the new incentives that would encourage industry to make products that last longer and are repairable, and to recycle and reuse materials. To be sure, people enjoy buying new things and often items are discarded not because they wear out but because a better camera or similar item has been developed. Yet much of what America buys does wear out quickly. The light bulb is only the most conspicuous example. Often items are discarded because there is no easy way to repair them.

Achieving greater durability in the products of industry would affect the demand for energy in a very fundamental way. If energy-intensive industrial products lasted twice as long as they do now, then annual energy requirements for those industries could be cut in half.

Changes in the manufacturing process to achieve such savings need not seriously disrupt the economy. Economic efficiency still would be the test but it would be measured over the life of the product and on the basis of the full cost to society. The objective would be to make products last as long as possible, easy to repair, and susceptible to recycling, or re-use. Innovation would not be stifled but redirected. New models just for the sake of change would no longer be in style. Mass-production techniques would not necessarily be abandoned, but jobs would of necessity be structured to provide the worker with enough variety and interest that he could take satisfaction in performing his tasks and in turning out a durable, high-quality product. In some instances this might mean completely mechanizing an operation and replacing the current human "robots" with machines overseen by men.

An important beginning point in moving toward more durable industrial products is to educate consumers to think more in terms of costs over the useful life of the items they buy. If all manufactured goods were accurately labeled to reveal their expected (or better still, guaranteed) life span and their cost per appropriate unit of time of useful service, we might see a new kind of product competition, i.e., winning a public reputation for quality. Another important step would be to put the tax laws to work to encourage durability and re-use. A tax incentive that increased with the length of time a product was guaranteed to last might be a very effective incentive. Regulations would be needed to require that autos and other products were built so that the materials they contain could be recycled. Equitable freight rates for transporting recycled materials could complete the package.

I believe that the concept of making quality and durability the hallmarks of industrial production should have strong appeal to a nation concerned with solving the shortage of energy and protecting the environment against damage from the rapid growth of energy-intensive industries, which are often also pollution-intensive. The concept should have appeal to consumers who get too little service and satisfaction for their money from present industrial products. Workers in the plants whose aspirations for human dignity are subverted by the mindless drudgery of their jobs have a direct and vital stake.

A common "bogey man" is that conserving energy will reduce employment. Here it is important to distinguish between planned conservation and a sudden unexpected shortage. Obviously if industry is suddenly deprived of energy it had counted on, production slows down and jobs are lost. But the gradual shift to homes that are better insulated, more efficient cars, and less energy intensive industrial production is no threat to employment. To be sure, a post-industrial society would no doubt mean fewer persons working in factories that produce goods, but it also means more persons who are better paid for providing skilled services. If worker productivity (as traditionally measured) were to decline in order to achieve quality and durability, the net effect would be fewer blue collar jobs in industry. But there would be an offsetting factor: Many more jobs would be created for individuals to maintain and repair the longer-lasting industrial products and to work in the industries which reclaimed and recycled materials. Another offsetting factor would be a shorter industrial work week; the material needs of society would be satisfied with fewer manhours of work per hour of useful service to the consumer. Furthermore, the new growth industries—education, the performing arts, farming, and others—would be more labor-intensive, creating additional and more rewarding jobs than those lost in the energy-intensive industries.

I do not mean to minimize the difficulties of achieving and maintaining full employment in the years ahead. We have fallen short of achieving this goal in times much less uncertain. Yet if we move in a planned and orderly manner, our shift to an economy in which energy is used more frugally poses no threat to employment opportunity for Americans.

I also recognize that the role of government—at least in terms of providing leadership—will need to be more sharply defined. New programs will have to be initiated, if the changes outlined in this chapter are to be achieved. But those who attempt to depict such a conservation-oriented society as coercive or stagnant are drawing a false picture. There is nothing coercive about a society in which buildings do not leak heat; in which people get to work by rapid transit, rather than by driving cars; where short trips are taken by high-speed rail rather than by driving or flying; and in which the products of industry are built to be durable and recyclable. It is a matter of reordering priorities and chaneling investments into energy-saving activities rather than more power plants and refineries. And such a society is not stagnant. Recreation, education, performing arts, farming, health care, research and development, and high technology would be some of its high-growth industries.

These conservation measures would be a part of a much broader restructuring of our economy and changing of our values, especially as these relate to material goods. The joy of buying something new would give way to the satisfaction of making something last. We could shift our preoccupation from possessions to intellectual and human concerns. We would have time to do more things with our hands, things we like to do. I do not foresee a hard or Spartan life in a slower, more frugal America. Indeed our energy requirements, though fairly stable, would be sufficient to provide decent housing, space conditioning the year round and basic conveniences such as a washing machine for everyone. But these material goods would move toward the background of our lives. Our satisfactions and growth would come more from our minds, spirits and relations with other people.

An essential ingredient of such a pattern of growth for America would be the building of new planned communities where jobs, housing and recreational activities are all near at hand. In these communities transportation would consist almost entirely of mass transit, walking and bicycling. Citizens would be relieved of two-hour auto commuting ordeals, now a dreary feature of suburban living. Recycling of wastes for fuel production and reuse in industry would be an integral part of the city plan. And the use of solar energy, waste heat, and other energy-conserving ideas would also be incorporated in the plan.

Some of these communities could be in the center of existing cities, others in new towns. Of course, a major government program will be needed if the new-communities concept is to become a reality. But no single public effort could save more energy and materials. The shape of present urban America, with expansion-ringing decaying central cities, is building enormous waste of time and energy and resources into the future. It will take many decades to reshape America, but certainly the time has come to make a beginning.

Industrial organizations will need to be remodeled to combine mass production with a new sense of concern for human beings in the plant and office.[263] If we do not do so, America may find that China will become the model for the Third World and perhaps the industrial nations as well. The Chinese not only have a frugal society where there is little waste of materials—they make lumber from sawdust, while we burn scrap—they have also apparently achieved a sense of purpose and participation in their industrial life that eludes us.

The American genius has been to create through a democratic form of government and financial incentives in the marketplace a free and open pursuit of happiness that no totalitarian government could hope to match. Now we must face the fact that too few Americans are genuinely happy in their jobs or with the quality of their lives. As columnist Bruce Biossat recently wrote:

> The plainest truth is that we don't know what to do with affluence. At least in American society, we have discovered that it produces an incredible sense of emptiness and boredom. It robs us of challenge, and we are finding that continuing challenge is essential to vibrant life.[264]

As we develop a more frugal society and aspire to new challenges we will not give up our material comforts; we will merely move such concerns away from the center of our lives. From any worldwide, long-term perspective this means that huge increases in the amount of energy will be required to achieve a desirable standard of living, and these levels must then be sustained. The future of mankind requires the development of renewable sources of energy. Only then will it be possible to aspire to an improving quality of life and flourishing civilization into the distant future.

There can be no more urgent task for mankind than to find, as rapidly as possible, alternatives to burning the limited fossil fuels on earth. It is not simply that an energy crisis is approaching. Any long-term projection of the need for hydrocarbons as an industrial raw material should convince us that their lavish and wasteful consumption as fuel must quickly cease. Petroleum, as a prime source of fertilizer, will be increasingly needed to prevent starvation on earth. And, in the long run, the fossil fuels may serve mankind better as a source of protein, than as a fuel for Sunday driving.

The world of energy has progressed steadily to more and more concentrated and sophisticated sources. The progression has been from wood to coal to petroleum and now to the atom. The next and perhaps ultimate step is to harness the sun directly or to control on earth the fusion reaction that occurs in the sun. Once such an inexhaustible source of energy is available, mankind would have the physical means to eradicate poverty. For it is clear that with inexhaustible and relatively clean and economical energy, mankind can satisfy the need for food and material comforts for all peoples who inhabit the earth.

My study of energy policy thus leads me to serious questions about America's present growth pattern as well as the urgent need for cleaner and more abundant sources for an energy-starved world. As government begins to grapple with these fundamental concerns, I am confident that the American people will respond vigorously to the challenge.

The energy crisis is an opportunity for greatness and service. We can serve not only the future generations of Americans but also people throughout the world. In coming decades they can be brought out of darkness and hopelessness into a global community where food and shelter are no longer a daily struggle and there is time for the pursuit and enjoyment of human values.

Notes

CHAPTER 2

1. Lewis Mumford, *Technics and Civilization*. New York: Harcourt, Brace and Company, 1934, pp. 157-158.

* 2. John W. Oliver, *The History of American Technology*. New York: The Ronald Press Company, 1956, pp. 331-332.

> When "Colonel" Edwin L. Drake developed the technique of drilling for oil in 1859, rather than waiting for it to seep through the ground and accumulate in pools, he opened up not only the hidden reservoirs of oil, but also a new industry and vast new fields for science and technology.
>
> The oil industry necessitated the building of new lines of railroads. It increased the production in iron and steel . . . It stimulated barge and shipbuilding . . . This abundant supply of petroleum developed an industry which had such a phenomenal growth that in ten years oil became the country's second greatest export, surpassed only by cotton. (Copyright © 1956 by Ronald Press Co., New York, pp. 331-332.)

3. Christopher Tugendhat, *Oil: The Biggest Business*. London: Eyre & Spottiswoode, 1968, p. 11.

4. Ida Tarbell, *The History of the Standard Oil Company*. New York: Macmillan Company, 1925, Vol. 2, pp. 283-288.

5. Sam M. Schurr, et. al., *Energy in the American Economy 1850-1975*, Resources for the Future, Baltimore, Maryland: Johns Hopkins University Press, 1960, p. 85.

* 6. Oliver, *The History of American Technology*. pp. 351, 337, 359.

> The opening of the Pearl Street Central Station in 1882 in New York City marked the beginning of the electrical age. Within four months the Edison Company was lighting more than five thousand lamps for two hundred and

> thirty customers, and the demand for lamps continued far beyond the company's ability to supply them. By 1890 it was supplying current to approximately twenty thousand lamps in many of the major office buildings in New York City.
>
> The electrical age introduced the second great power age in our nation's history, steam being the first. By 1900 the basic types of motors had been invented. The alternating current had become the system for generating, transmitting, and utilizing electric power. The transformer, high voltages, the induction motor, the long distance transmission line made possible the rapid spread of electric light and power. (Copyright © 1956 by Ronald Press Co., New York, pp. 331-332.)

7. Hans H. Landsberg and Sam M. Shurr, *Energy in the United States*, Sources, Uses & Policy Issues. New York: Random House, 1960, p. 52.

Mumford, *Technics and Civilization*, p. 224.

8. Tugendhat, *Oil: The Biggest Business*, pp. 82-96.

9. One of the early uses for coal was as a manufactured gas for the gaslights and for cooking and heating water. The first gas company was organized in Baltimore, Maryland, in 1816, but gas was very expensive in those days ($10 per thousand cubic feet, as compared to about $1.50 today) and therefore not widely used.

For further reading, see Paul J. Garfield, Ph.D., and Wallace F. Lovejoy, Ph.D., *Public Utility Economics*. Englewood Cliffs, New Jersey: Prentice-Hall, Inc., 1964, and Louis Stotz and Alexander Jamison, *History of the Gas Industry*. New York: Stettiner Brothers, 1938.

10. Schurr, et. al., *Energy in the American Economy 1850-1975*, pp. 126-127.

> Prior to the late 1920's the transportation of natural gas over a distance of 200 or at most 300 miles was considered an outstanding accomplishment. But toward the end of the 1920's, the use of conventional screw couplings in connecting the pipe were replaced by welded pipe joints. Seamless pipes of larger diameter began to be manufactured from steel of great tensile strength which permitted high transmission pressures. Thus greater use could be made of the high natural pressure existing in some gas fields. Through improvements in compressor stations, which restore the pressure as the gas moves along, long-distance lines could transport several times the volume of the gas piped without recompression. These improvements and the use of heavy power equipment for the laying of pipe extended the

range over which natural gas could be transported easily and economically to 1,000 miles by the mid-1930's.

11. Garfield and Lovejoy, *Public Utility Economics*, p. 167.

CHAPTER 3

12. The btu (British thermal unit, which is the quantity of heat required to raise the temperature of one pound of water 1 degree Fahrenheit) is used here as a common standard of measurement for the different forms of energy.

13. For a valuable discussion of the energy/GNP ratio concept see Schurr, et. al., *Energy in the American Economy 1850-1975*, pp. 157-190 and "Energy Consumption and Gross National Product in the United States," National Economic Research Associates, Inc., Washington, D.C., March 1971.

14. Approximately 95% of all homes in the United States are wired for electricity. Oran L. Culberson, *The Consumption of Electricity in the United States*. Oak Ridge: Oak Ridge National Laboratory, June 1971, p. 10.

15. *Patterns of Energy Consumption in the United States*, prepared for Energy Policy Staff, Office of Science and Technology, Office of the President by Stanford Research Institute, Menlo Park, California, November 1971.

16. *Ibid.*

17. *The Economy, Energy, and the Environment*, A Background Study, prepared for the use of the Joint Economic Committee, Congress of the United States, by the Environmental Policy Division, Legislative Reference Service, Library of Congress, September 1, 1970, p. 1.

18. *Patterns of Energy Consumption in the United States*, p. 88 t. sq. Eight specific industries or products alone consumed as estimated 14.5% of the total energy consumption in 1968. Their share—in percent of total energy consumption—has been estimated as follows:

Iron and Steel	4.5%
Petroleum Refining	3.8%
Paper Board	1.7%
Petrochemical Feedstock	1.6%
Aluminum	.9%
Cement	.7%
Ammonia	.7%
Ferrous foundries	.7%

19. "A Review and Comparison of Selected United States Energy Forecasts," prepared for the Executive Office of the President, Office of Science and Technology, Energy Policy Staff by Pacific Northwest Laboratories of Battelle Memorial Institute, December 1969; also, "Energy Demand Studies—An Analysis and Appraisal," prepared for the use of the Committee on Interior and Insular Affairs of the U.S. House of Representatives, 92nd Congress, September 1972. Both studies give valuable detailed critiques of projections made by research organizations, government agencies and the different branches of the industry.

20. The results of existing projections are similar because they are based on the same history and most make similar assumptions about the future growth of population and GNP.

21. Department of Commerce, Bureau of the Census; *Current Population Reports*, Series P-25, No. 448, Series E.

22. See Duane Chapman, Timothy Tyrell and Timothy Mount, "Electricity Demand Growth and the Energy Crisis," *Science*, Vol. 178, Number 4062, November 17, 1972, pp. 703-708.

Timothy Tyrell, "Projections of Electricity Demand," Oak Ridge National Laboratory, November 1973.

Kent P. Anderson, "Residential Energy Use: An Econometric Analysis," prepared by the RAND Corp. for the National Science Foundation, October, 1973.

J. W. Wilson, *Quarterly Review of Economics and Business, II*, No. 1, Spring, 1971.

P. W. MacAvoy, *Economic Strategy for Developing Nuclear Breeder Reactors*. Cambridge, Mass.: MIT Press, 1969.

R. Halvorsen, *Sierra Club Conference on Power and Public Policy*. Burlington, Vermont: Public Resources, Inc., 1972.

For studies of natural gas see J. E. Draper, "Reviews of Seven Gas Demand Studies Utilizing Econometric Techniques," appearing as appendix to Chapter 10; Keith C. Brown, ed., *Regulation of the Natural Gas Producing Industry*, Baltimore: Resources for the Future, Johns Hopkins University Press, 1972.

23. Much of the statistical analysis of this section is derived from Joel Darmstadter, et. al., *Energy in the World Economy* and Joel Darmstadter, "Energy Consumption: Trends and Patterns," appendix to *Energy, Economic Growth and the Environment,* Sam H. Schurr ed., both books published for Resources for the Future, by Johns Hopkins University Press, 1971, 1972.

CHAPTER 4

24. "An Economic Profile of Mainland China," Studies Pre-

pared for the Joint Economic Committee, Congress of the United States, 90th Congress, 1st Session.

25. "Environmental Quality," *Congressional Record*, A Study by the Congressional Action Fund, September 19, 1972, p. 15, 237.

26. *Ibid.*, p. 15, 236.

> Fuel combustion in stationary sources such as home heating and power production produces 73% of sulfur oxides and 42% of nitrogen oxides . . ., while motor vehicles produce 73.7% of all carbon monoxide and 52.9% of hydrocarbons. . . . Industrial processes produce 40.9% of the particulates and 22.5% of the sulfur oxides.

See also Lester Lave, "The Health Hazards of Electricity Generation," Carnegie-Mellon University, December 1971.

27. Marshall I. Goldman, ed., *Controlling Pollution, The Economics of a Cleaner America*. Englewood Cliffs, New Jersey: Prentice-Hall, Inc., 1967, p. 52.

28. See "Air Pollution Episodes," A Guide for Health Departments and Physicians, Arlan A. Cohen, M.D., et. al., Environmental Protection Agency, HSMHA Health Reports, June 1971, Vol. 86, No. 6.

29. Ronald G. Ridker, *Economic Costs of Air Pollution*. New York: Frederick A. Praeger, 1967, pp. 30-56.

30. *Particulate Polycyclic Organic Matter, Biological Effects of Atmospheric Pollutants*. Washington, D.C.: National Academy of Sciences, 1972, pp. 13-35.

31. *Electric Power and the Environment*, U.S. Office of Science and Technology, The Energy Policy Staff, August 1970, p. 3.

32. Anthony Wolff, "Showdown at Four Corners," *Saturday Review of Society*, June 3, 1972, pp. 29-41.

> In the Four Corners area, once celebrated for the Iberian clarity of its sky, visibility has been drastically reduced. One-hundred-mile vistas have been curtained off: Shiprock Peak, the world-famous, 700-foot-high monolith twenty miles to the west of the plant, is veiled in haze. From forty miles and more to the east, high on the slopes of the Continental Divide, the airborne soot can be seen trailing south along the spine of Hogback Mountain. By plane, the pall has been detected over an area of thousands of square miles, and the concentrated plume from the plant's stacks has become a landmark beacon for high-flying pilots. In 1966 the Gemini XII astronauts reported

that the plume was the only human artifact visible from outer space.

33. *National Air Quality Levels and Trends in Total Suspended Particulates and Sulphur Dioxide Determined by Data in the National Surveillance Network*, EPA, April 1973.

34. *Particulate Polycyclic Organic Matter, Biologic Effects of Atmospheric Pollutants*, pp. 237-246.

35. Dr. Lester Machta, "Prediction of Carbon Dioxide in the Atmosphere," presented at the Brookhaven Symposium in Biology, "Carbon and and Biosphere," May 16-18, 1972.

36. Richard H. Wagner, *Environment and Man*. New York: W. W. Norton and Company, 1971, pp. 190-192.

37. *Restoring the Quality of Our Environment*, Report of the Environmental Pollution Panel of the President's Science Advisory Committee. Washington, D.C.: Government Printing Office, November 1965, pp. 112-128; *Environmental Quality*. Washington, D.C.: Government Printing Office, August 1970, pp. 95-97.

38. *Man's Impact on the Global Environment*. Cambridge, Mass.: The MIT Press, 1970, p. 55.

39. Claire Sterling, "Doomsday Prophecies about Climate are Eased Some," *Washington Post*, January 8, 1972.

40. David A. Andelman, "20% Rise Feared in Carbon Dioxide," *The New York Times*, May 17, 1972.

41. Theodore L. Brown, *Energy and the Environment*. Columbus, Ohio: Charles E. Merrill Publishing Company, 1971, pp. 61-71.

Helmut E. Landsberg, "Man-Made Climatic Changes," *Science*, December 18, 1970, Volume 170, Number 3964, pp. 1,265-1,273.

42. *Man's Impact on the Global Environment*, p. 64.

43. *Environmental Conservation*, The Oil and Gas Industries, A National Petroleum Council Study, 1971, Volume 1, p. 63.

44. *Ibid*. p. 59.

45. U.S. Department of Interior, Final Environmental Impact Statement on Outer Continental Shelf Oil and Gas, General Lease Sale off Eastern Louisiana, originally scheduled for December 21, 1971, p. 31.

46. *Annual Report of the President to the Congress on Marine Resources and Engineering Development, Marine Science Affairs, Selecting Priority Programs*, April 1970, p. 65.

47. Max Blumer, "A Small Oil Spill," *Environment*, March 1971, Volume 13, Number 2, p. 11.

See also: Testimony of Max Blumer before the Subcommittee on Air and Water Pollution, Senate Committee on Public Works, Machias, Maine, June 30, 1970. Testimony of Max Blumer before the Antitrust and Monopoly Subcommittee, United States Senate, August 4-15, 1970. For an opposing view see Dr. Dale Straughan, *Summary of Biological Effects of Oil Pollution in the Santa Barbara Channel*, Part I, Allan Hancock Foundation, University of Southern California, February 17, 1971, and for criticism of Dr. Straughan's work see Dr. J. H. Connell, *Report for Australian Royal Commission on Oil Exploitation of the Great Barrier Reef*, quoted in Robert Eaton, *Black Tide, The Santa Barbara Oil Spill and its Consequences*. New York: Delacorte Press, 1972, pp. 255-256.

48. Blumer, "A Small Oil Spill," p. 5.

49. According to the House Committee on Government Operations, which held hearings in December 1971, there were 8,496 oil spills during 1971. The Coast Guard, which is authorized to enforce provisions of the water pollution laws and the 1899 Rivers and Harbors Act, "Referred only 372 cases to U.S. attorneys for prosecution under the Refuse Act and assessed civil penalties in only 145 cases." Unless these laws are more rigorously enforced and fines are assessed, there can be little hope of stemming the flow of oily pollutants into our coastal and estuarian waters. "Protecting America's Estuaries," House Committee on Government Operations, December, 1971.

50. Malcolm F. Baldwin, *Public Policy on Oil, An Ecological Perspective*. Washington, D.C.: The Conservation Foundation, 1971, p. 26.

51. *Petroleum Today*, published by American Petroleum Institute, Summer, 1971, p. 11.

52. Frederick A. Moritz, "Giant Tankers to Mean Bigger Oil Spills?" *Christian Science Monitor*, September 19, 1972.

53. *The Oil Spill Problem*, First Report of the President's Panel on Oil Spills, Office of Science and Technology, 1969.

Offshore Mineral Resources, A Challenge and an Opportunity, Second Report of the President's Panel on Oil Spills, Office of Science and Technology, 1969.

54. *Environmental Conservation*, p. 55.

55. A pipeline explosion on March 4, 1965, killed 17 persons and destroyed five homes near Natchitoches, Louisiana. A survey covering a fifteen-year period through 1965 revealed 64 deaths and

225 injuries from pipeline failures. *Federal Power Commission*, 1966 Annual Report, p. 135.

56. Senator Nelson, "Strip Mining," *Congressional Record*, April 5, 1971.

> Already, an estimated 1.8 million acres have been disturbed by strip mining in this country. And at presently accelerating rates, the figure will reach 5 million acres, an area about the size of New Jersey, by 1980, the Interior Department estimates.
>
> Yet the Department finds that only 56,000 acres have thus far been reclaimed after strip mining.

57. James Branscome, "Appalachia—Like the Flayed Back of a Man," *The New York Times Magazine*, December 12, 1971.

58. Ben A. Franklin, "Coal Rush Is on as Strip Mining Spreads into West," *The New York Times*, August 21, 1971.

59. John F. Stacks, *Stripping, The Surface Mining of America*. San Francisco: The Sierra Club, 1971, pp. 81-82.

60. Carl Bagge, "Coal, an Overlooked Energy Source," The 1972 Gabrielson Lecture delivered at Colby College, Waterville, Maine, March 2, 1972.

See statements of Edward Phelps, president of Peabody Coal Co., Ralph Hatch, president of Hanna Coal Co., and Paul Morton, president of Cornelton Coal Co. in "Regulation of Strip Mining," Hearings before the Subcommittee on Mines and Mining of the Committee on Interior and Insular Affairs, House of Representatives, 92nd Congress, 1st Session, November 29, 1971.

61. The General Accounting Office uncovered many rule violations by both the Department of Interior's Bureau of Land Management and the Bureau of Indian Affairs in their lack of enforcement of environmental protection laws. This resulted in the failure of many mining companies to make effective provisions for reclaiming the strip-mined land.

For further details, see *Administration of Regulations for Sulphur Exploration, Mining, and Reclamation of Public and Indian Coal Lands*, A report by the Comptroller General of the United States, General Accounting Office, August 10, 1972.

62. See E. A. Nephew, "Surface Mining and Land Reclamation in Germany," Oak Ridge National Laboratory, May 1972, pp. 56-57.

63. "Surface Mining," Hearings before the Subcommittee on Minerals, Materials and Fuels, of the Committee on Interior and Insular Affairs, United States Senate, 92nd Congress, 1st Session, November-December 1971.

"Regulation of Strip Mining," Hearings before the Subcommittee on Mines and Mining of the Committee on Interior and Insular Affairs, House of Representatives, 92nd Congress, 1st Session, September, October and November 1971.

64. *Testimony of Russell E. Train, Chairman, Council on Environmental Quality*, before Subcommittee on Minerals and Fuels, Committee on Interior and Insular Affairs, United States Senate, 92nd Congress, 1st Session, November 16, 1971, pp. 141-142.

65. *Underground Coal Mining in the United States, A Study of Research and Development Programs*, TRW Systems Group, June 1971, Appendix C, p. C-3.

66. "Clean Water for the 1970's," A Status Report, The Federal Water Quality Administration, June 1970.

67. Harry Perry, "Environmental Aspects of Coal Mining," in David A. Berkowitz and Arthur M. Squires, eds., *Power Generation and Environmental Change*. Cambridge, Mass.: The MIT Press, 1971, p. 334.

68. *Ibid.*, pp. 324-325.

> A total of about 1.3 billion tons of bituminous rejects from cleaning plants have been produced to date. An additional unestimated quantity of rejects from hard preparation has been produced. If this refuse has an average density of approximately 1 ton per cubic yard and the average height of the refuse banks is 75 feet, then the waste material from bituminous coal preparation plants would cover 11,000 acres.

69. Michael Fortune, "Environmental Consequences of Extracting Coal," Electric Power Consumption and Human Welfare, Power Study Group of the Committee on Environmental Alterations of the American Association for the Advancement of Science, pre-publication manuscript.

> Underground fires are difficult to extinguish: One has burned at New Straitsville, Ohio, since 1884, despite over one million dollars spent trying to put it out; $5-1/2 million has been spent in vain on a fire in Scranton, Pennsylvania. Again, foresight is the only cure—65% of these blazes start in garbage heaps in abandoned strip mines.

70. Ben A. Franklin, "U.S. Warned West Virginia in 1967 that 30 Coal-Waste Piles Were Unstable," *The New York Times*, March 3, 1972.

71. *Washington Post*, October 12, 1972.

> Seventeen Southeastern Kentucky coal miners escaped in-

jury Tuesday when a rupture in a coal-waste pond bottom sent torrents of black water gushing through the Bethlehem Steel Corporation mine near here.

72. George Vecsey, "Mine Safety Effort Draws Rising Criticism," *The New York Times*, September 9, 1972.

73. See Henry N. Doyle, "An Action Program for the Prevention of Coal Pneumoconiosis," Bureau of Mines, Department of Interior, November 1968.

L. E. Kerr, "Coal Workers and Pneumoconiosis," Archives of *Environmental Health*, 16:4, April 1968, p. 579.

W. S. Lainhart, et. al., *Pneumoconiosis for Appalachian Bituminous Coal Miners*, U.S. Public Health Service publication, No. 2000, 1969.

74. *Underground Coal Mining in the United States*, Ch. 4, pp. 3-12.

75. William Grieder, *Washington Post*, December 20, 1971, p. A14.

76. *Steam Power Plant Site Selection*, A Report Sponsored by the Energy Policy Staff, Office of Science and Technology, December 1968.

Electric Power and the Environment, A Report Sponsored by the Energy Policy Staff, Office of Science and Technology, August 1970.

"Powerplant Siting and Environmental Protection," Hearing before the Subcommittee on Communications and Power of the Committee on Interstate and Foreign Commerce, House of Representatives, 92nd Congress, 1st Session, 1971.

"Problems of Electrical Power Production in the Southwest," Report of the Committee on Interior and Insular Affairs, U.S. Senate, 92nd Congress, 1st Session, 1972.

Neil Fabricant and Robert M. Hallman, *Toward a Rational Power Policy, Energy, Politics and Pollution*. New York: George Braziller, Inc., 1971.

77. Fabricant and Hallman, *Toward A Rational Power Policy, Energy, Politics, and Pollution*, pp. 149-153.

78. William H. Steigelmann, "Alternative Technologies for Discharging Waste Heat," *Power Generation and Environmental Change*, p. 399.

79. See Dorothy Nelkin, *Nuclear Power and Its Critics—The Cayuga Lake Controversy*. Ithaca, N.Y.: Cornell University Press, 1971.

80. *Considerations Affecting Steam Power Plant Site Selection*,

U.S. Office of Science and Technology, The Energy Policy Staff, December 1968, p. 39.

81. *The 1970 National Power Survey*, Federal Power Commission, December 1971, Part 1, Chapter 10, pp. 3-12.

82. Steigelmann, *op. cit.*, p. 405.

> Although the dry cooling tower is presently not competitive with other methods and systems for accomplishing cycle heat rejection and therefore is not used unless water for cooling purposes is unavailable, considerable interest is being shown in the system by virtually all electric utility organizations. This interest results from a combination of at least two factors: First, the appeal of a system that does not raise thermal pollution issues, does not require water makeup, and does not create a moisture-laden plume; this can mean that a site otherwise unacceptable can be used, that many disputes with the public and regulatory authorities at hearings can be avoided, and that expansive programs to monitor the aquatic environment can be avoided. Second, since large-scale dry cooling systems for power plants have not yet been built, there is the strong possibility that technological improvements will lead to significantly lower costs.

83. *Testimony of William O. Ruckelshaus, Administrator, Environmental Protection Agency*, before the Joint Committee on Atomic Energy, U.S. Congress, 92nd Congress, 1st Session, 1971, Part 1, pp. 229-230.

84. Anthony Ripley, "Radioactive Building Sand Stirs Dispute," *The New York Times*, September 27, 1971.
———"Infants and Radioactive Sands: Small-Town Doctor Wins Fight," *The New York Times*, October 3, 1971.
———"City in Colorado Awakens to Scope of Radioactive Waste Problems," *The New York Times*, October 4, 1971.
———"Radioactive Sands Linked to Higher Death Rate," *The New York Times*, October 28, 1971.
Sheldon Novick, *The Careless Atom*. New York: Dell Publishing Company, 1969, pp. 134-136.
"Pupils Sit Over Landfill of Uranium Waste," *The Washington Evening Star*, October 19, 1972.

85. *Testimony of Dr. Reuben G. Gustavson, vice president and Dean of the Faculties, University of Chicago*, before Special Committee on U.S. Atomic Energy, U.S. Senate, 79th Congress, 2nd Session, January 22, 1946.

86. Merril Eisenbud, "Standards of Radiation Protection and

Their Implications for the Public's Health," in Harry Foreman, M.D., ed., *Nuclear Power and the Public*. Minneapolis: University of Minnesota Press, 1970.

87. Federal Register, Vol. 36, No. 111, Atomic Energy Commission, Proposed Rule Making, June 7, 1971, p. 11,114.

Among the results the AEC expects to achieve through conformance with the guides on design objectives are the following:

> 1. Provide reasonable assurance that annual exposures to individuals living near the boundary of a site where one or more light-water-cooled nuclear power reactors are located, from radioactivity released in either liquid or gaseous effluents from all such reactors, will generally be less than about 5 percent of average exposures from actual background radiation. This level of exposure is about 1 percent of Federal radiation protection guides for individual members of the public.
>
> 2. Provide reasonable assurance that annual exposures to sizable population groups from radioactivity released in either liquid or gaseous effluents from all light-water-cooled nuclear power reactors on all sites in the United States for the foreseeable future will generally be less than about 1 percent of exposure from natural background radiation. This level of exposure is also less than 1 percent of Federal radiation protection guides for the average population dose.

88. Arthur R. Tamplin, "Issues in the Radiation Controversy," *Science and Public Affairs, Bulletin of the Atomic Scientists*, September 1971, p. 26.

89. *Estimates of Ionizing Radiation Doses in the United States 1960-2000*, U.S. Environmental Protection Agency, Office of Radiation Programs, Division of Criteria and Standards, August 1972.

90. Robert Gillette, "Nuclear Reactor Safety: At the AEC the Way of the Dissenter is Hard," *Science*, Vol. 176, May 5, 1972, p. 492.

> The argument over ECCS (the emergency core cooling system) is neither academic nor trivial. Should a reactor's searingly hot core run dry, the ECCS is supposed to reflood it with water within seconds after the leak occurs. Should the ECCS fail—or even hesitate for long—the core could melt and ensuing steam explosions could scatter its radioactive contents over a wide area. The indications are that existing designs of backup cooling systems might not adequately reflood a reactor after a major leak.

91. Presently, under the law, the Price-Anderson Act passed in 1957, the federal government's maximum share of the liability for damage from a nuclear power plant is $478 million while private industry insures the plant for $82 million dollars in damages.

92. Joel A. Snow, "Radioactive Waste from Reactors," *Scientist and Citizen*, May 1967, p. 91.

93. Gene Bryerton, *Nuclear Dilemma*. New York: A Friends of the Earth/Ballantine Book, 1970, p. 55.

94. According to a study of 183 tanks at the Hanford, Savannah River and Idaho waste disposal sites, Belter and Pearce found nine instances of tank failure. They concluded that a tank may only last several decades. W. S. Belter and D. W. Pearce, "Radioactive Waste Management," *Reactor Technology*, Atomic Energy Commission, January 1965, p. 203.

95. Plan for the Management of AEC-Generated Radioactive Wastes, Wash-1202, U.S. Atomic Energy Commission, Division of Waste Management and Transportation, January 1972, p. 20.

96. For a comprehensive, recent discussion of the problem see Mason Willrich, *Global Politics of Nuclear Energy*. New York: Praeger Publishers, 1971.

Mason Willrich, ed., *Civil Nuclear Power and International Security*. New York: Praeger Publishers, 1971, pp. 3-4.

> By 1980 about 300,000-450,000 kilograms of plutonium will be accumulated in civil nuclear power programs throughout the world. Almost half of it will be located in the United States, but thousands of kilograms will be available in the near future in most of the key nations which have thus far refrained from acquiring nuclear weapons.
>
> Between 5 and 10 kilograms of plutonium or less than 20 kilograms of fully enriched uranium is needed for a bomb which will destroy a medium-size city. The latent threat to the security of various nations resulting from the widespread availability of fissionable materials and the technology to produce these materials will clearly be very large.

97. Willrich, *Civil Nuclear Power and International Security*, p. 93.

98. Theodore B. Taylor, "The Need for National and International Systems to Provide Physical Security for Fissionable Materials," paper presented before the American Association for the Advancement of Science, December 28, 1971.

> . . . present civilian inventories of fissionable materials in forms suitable for use in nuclear explosives (without requiring extensive material processing facilities) are of the order of several thousand kilograms in the United States; non-U.S. inventories are probably somewhat smaller. Both are increasing rapidly.
>
> . . . the knowledge, materials, and facilities required to make nuclear explosives with yields at least as high as the equivalent of a few tens of tons of high explosive, given the required fissionable materials, are distributed worldwide. They could be assembled by groups of people with resources available to practically any country in the world. There is no longer any secret of the "Poor Man's Atomic Bomb."

CHAPTER 5

99. *Energy R&D and National Progress: Findings and Conclusions*, Prepared for the Interdepartmental Energy Study by the Energy Study Group under the direction of Ali Bulent Cambel, September 1966, p. 4.

100. These are oil, natural gas, coal and hydroelectric power. In later chapters we shall examine other sources that could be made available such as shale oil, geothermal energy, solid and organic wastes, solar energy, nuclear breeders and fusion power.

101. In 1925 the United States accounted for more than 70 percent of the total oil production in the world. The U.S. production was not exceeded until 1970, when the Middle East became the world leader. Currently, domestic oil production is about 11 million barrels per day, while the total output from Middle Eastern nations is around 17 million barrels per day.

102. P. K. Theobald, S. P. Schweinfurth, and D. C. Duncan, *Energy Resources of the United States*. Washington, D.C.: Geological Survey Circular 650, p. 1, and *United States Energy Through the Year 2000*. Washington, D.C.: U.S. Department of Interior, 1972, p. 57.

103. *The World Petroleum Market*, page 32, lists various recovery percentages obtainable from oil fields using various types of drive mechanism for production equipment. The lowest median percentage is 18, and the high is 51. These figures also vary with the type of oil-bearing formation and the size of the reservoir. M.A. Adelman, *The World Petroleum Market*, published for Resources for the Future, Inc., London and Baltimore: Johns Hopkins University Press, 1972, p. 32.

104. M. K. Hubbert, "Nuclear Energy and the Fossil Fuels," in *Drilling and Practice—1956*. New York: American Petroleum Institute, 1957.

105. T. A. Hendricks, "Resources of Oil, Gas, and Natural Gas Liquids in the United States and the World," U.S. Geological Survey Circular 522, 1965.

106. *Report on Estimates of Additional Recoverable Reserves of Oil and Gas for the United States and Canada,* Degoyler and MacNaughton, June 30, 1969.

107. *Federal Power Commission News*, Volume 6, Number 17. Week ended April 27, 1973, p. 4.

108. It must be noted that the industry recovers an average of only 30 percent of the Oil Resources In-Place. National Petroleum Council, *U.S. Energy Outlook: Full Report of the National Petroleum Council's Committee on U.S. Energy Outlook*, Washington, D.C., December 1972, p. 25.

109. In 1950, gas represented about 18 percent of total energy consumption. By 1960 it climbed to 28.2 percent of the total and in 1970 it reached 32.8 percent. *Energy Research Needs*, A Report to the Nation prepared by Resources For the Future, Inc. in cooperation with the M.I.T. Environmental Laboratory, October 1971, pp. 1-16.

110. Robert Sherril, "Energy Crisis! The Industry's Fright Campaign," *The Nation*, June 26, 1972.

111. *United States Energy Through the Year 2000*. Washington, D.C.: U.S. Department of Interior, December 1972, pp. 55-56, and *Reserves of Crude Oil, Natural Gas Liquids, and Natural Gas in the United States and Canada as of December 31, 1971*, American Gas Institute, American Petroleum Institute, Canadian Petroleum Institute, May 1972, p. 113.

112. *U.S. Energy Outlook: Full Report of the National Petroleum Council's Committee on U.S. Energy Outlook*, National Petroleum Council. Washington, D.C., Dec. 1972.

113. *Reserves of Crude Oil, Natural Gas Liquids, and Natural Gas in the United States and Canada*, *op. cit.*, p. 124.

114. Ralph E. Lapp, "We're Running Out of Gas," *The New York Times Magazine*, March 19, 1972.

> The new development is a sort of floating thermos bottle, a super-cold, double-hulled, refrigerated ship . . . The vessel is capable of maintaining a temperature of minus 259° F.—the point at which natural gas turns liquid—in its huge tanks, and can carry a billion cubic feet of natural gas.
>
> All told, synthetic gas and imported gas—coming by pipeline from the north and by ship from overseas—promise to add 7 trillion or 8 trillion cubic feet to the

United States' supply each year by the mid-eighties. According to our best estimates now, this and the conventional domestic supply will still leave an unsatisfied demand of some 13.7 trillion cubic feet by 1985.

115. This accounts for coal in either known or likely deposits less than 3,000 feet deep and in seams recoverable under current technology and economics. About half this amount, however, is in the sub-bituminous grades of coal which are dirtier burning and have less heating value than the higher grades. *The Economy, Energy, and the Environment, A Background Study*. Joint Committee Print, by the Environmental Policy Division, Legislation Reference Service, Library of Congress, printed for the use of the Joint Economic Committee, September 1, 1970. (Note: data for this study derived from U.S. Geological Survey estimates.)

116. P. K. Theobald, S. P. Schweinfurth, D. C. Duncan, "Energy Resources of the United States," Washington, D.C.: U.S. Geological Survey Circular 650, p. 1.

117. *Bituminous Coal Data*, 1972 Edition, National Coal Association.

118. *Statistical Abstract of the United States*, 1972.

119. *Bituminous Coal Data*, 1971 Edition, National Coal Association.

120. Even with the low efficiency in current nuclear plants, one pound of uranium does the work of 8 tons of coal. The energy potential of a pound of uranium is therefore 16,000 times as great as in coal. *The Nuclear Industry 1971*, U.S. Atomic Energy Commission, pp. 13-21.

121. AN ESTIMATE OF DOMESTIC URANIUM (AS U_3O_8) RESOURCES AS A FUNCTION OF PRICE

Price Range (*up to $/lb.* U_3O_8)	Cumulative		
	Reasonably Assured	Estimated Additional	TOTAL (*in tons* U_3O_8)
$10	390,000	680,000	1,070,000
$30	750,000	1,660,000	2,410,000
$50	4,750,000	3,660,000	8,410,000
$100	8,750,000	8,660,000	17,410,000
$200	Possibly $2 billion land and $4 billion in sea water.		

From *Energy Research Needs*, p. 73.

122. *The 1970 National Power Survey*, Federal Power Commission. Washington, D.C.: U.S. Government Printing Office, December 1971, Part 1, p. 6, and *United States Energy, A Summary Review*, U.S. Department of the Interior, January 1972, p. 22.

123. Report to the Secretary of the Interior of the Advisory Committee on Energy, June 30, 1971, p. 14.

> . . . There will be enough Uranium-238 in storage as a by-product from uranium enriching operations for light water reactors by the year 2000 to sustain breeder reactors then in operation for over five hundred years. When this supply is exhausted, the power then can be produced from uranium mined from the ground at less than $50/lb. which will last about 1200 years.

124. *The 1970 National Power Survey*, p. I-7-1.

> At the end of 1970, conventional hydroelectric capacity totaled 52,323 megawatts.
>
> By 1990, conventional hydroelectric capacity is expected to total approximately 82,000 megawatts.
>
> The projected total capacity in 1990 is approximately 1,260,000 megawatts, giving hydroelectric capacity 6.5% of total electric power and assuming electricity is still approximately 25% of total energy, only 1.6% of total energy.

125. *Ibid.*

> They [hydroelectric power plants] occupy large areas of land . . . and cause short- or long-term changes in stream regimens, including such items as reservoir drawdown, so they are often strenuously opposed on esthetic or ecological grounds.

126. The reservoir, however, may well eventually fill with silt. So that a hydro-plant is not necessarily an everlasting source of energy.

127. *BP Statistical Review of the World Oil Industry 1970*, The British Petroleum Company Limited, London, p. 5, and Frank J. Gardner, "Price, Nationalization Jitters Plague International Oil World," *Oil & Gas Journal*, December 27, 1971, p. 72.

128. John P. Albers, et. al., "Summary Petroleum and Selected Mineral Statistics for 120 Countries, Including Off-Shore Areas," Geological Survey Professional Paper 817, Washington, D.C., 1973.

129. Gardner, "Price, Nationalization Jitters Plague International Oil World," p. 73.

130. In 1972, Canada exported about 1,200,000 b/d (barrels per

day) to the U.S. while Venezuela exported another 2,300,000 b/d.

131. Venezuela produced an estimated total of 3,285,000 b/d in 1972.

132. *United States Energy, A Summary Review*, U.S. Department of the Interior, January 1972, p. 53, and *Energy Research Needs*, A Report to the National Science Foundation, p. IV-32.

133. M. K. Hubbert, "The Energy Resources of the Earth," *Scientific American*, September 1971, p. 65, and Theodore A. Hendricks, "Resources of Oil, Gas, and Natural Gas Liquids in the United States and the World," *Geological Survey* Circular 522, Washington, D.C., 1965, p. 17.

134. Hendricks, "Resources of Oil, Gas and Natural Gas Liquids in the United States and the World," p. 17.

135. *Report on Estimates of Additional Recoverable Reserves of Oil and Gas for the United States and Canada*, DeGolyer and MacNaughton, Dallas, Texas, June 30, 1969, p. 24.

136. Lapp, "We're Running Out of Gas"

> The well-head price of Algerian gas is about 4 cents a thousand cubic feet. After it's piped to the coast and liquified, the price is about 35 cents. When the [tanker] *Descartes* delivers it to the special storage facility at Everett, a thousand cubic feet is worth about 71 cents. And when it's been converted back into a gas, it wholesales for around $1.15.

137. Note that the recent agreement to provide 365 billion cubic feet of natural gas annually for 25 years between Sonatrach (the Algerian state gas and oil monopoly) and El Paso was signed in 1969. It took until 1973 before the FPC gave their final approval to this $1.5 billion project.

138. *Energy in the World Economy*, pp. 46-47.

COAL RESOURCES DETERMINED BY MAPPING AND EXPLORATION

United States	1,420 billion tons
U.S.S.R.	5,900 billion tons
Asia	450 billion tons
World	8,610 billion tons

139. *An Economic Profile of Mainland China*, Studies Prepared for the Joint Economic Committee, Congress of the United States, 90th Congress, 1st Session, Volume 1: General Economic Setting,

The Economic Sectors, Washington, D.C.: U.S. Government Printing Office, February 1967, pp. 302-304.

140. *The Uranium Enrichment Service Criteria*, established on December 23, 1966, by the Atomic Energy Commission.

> This order was issued to protect and maintain a viable domestic industry. It exclusively forbids the enrichment of imported uranium. In other words, uranium that was imported from Canada could not be enriched in the United States and used in New York. This prohibition thus created a *de facto* embargo on foreign uranium.

CHAPTER 6

141. Tugendhat, *Oil: The Biggest Business*, p. 118.

142. William L. Shirer, *The Rise and Fall of the Third Reich*. New York: Simon and Schuster, 1960, p. 1,097.

Albert Speer, *Inside the Third Reich*. New York: Macmillan, 1970, p. 414.

143. Cabinet Task Force on Oil Import Control, *The Oil Import Question*. Washington, D.C.: U.S. Government Printing Office, 1970, p. 36.

144. *U.S. Energy Outlook—An Initial Appraisal by the Energy Demand Task Group 1971-85*. National Petroleum Council's Committee on U.S. Energy Outlook.

145. Statement by the Honorable George P. Schultz, Secretary of the Treasury, at the American Bankers Association International Monetary Conference, Paris, France, June 6, 1973.

CHAPTER 7

146. For example, the cost of purchased fuels and electricity as a percentage of value added in manufacturing industry was only 3.01 in 1967 as reported in U.S. Department of Commerce, "Fuels and Electric Energy used in Manufacturing." Preliminary Report 1967 Census of Manufactures Series MC67 (p)-7, July 1969.

147. *Public Power*, July-August, 1972, p. 13. Actually the rates in the sliding scale increased by 12 percent from 1950 to 1968, as contrasted to a 44 percent increase in the general Consumer Price Index. The impact of the sliding scale on increased consumption more than offset the rate increases with the resulting major reduction in the average price.

148. The average per-capita consumption of electric power by consumers served by investor-owned companies, which serve about 30 percent of the public, increased from 1,770 kwh in 1950 to 6,700 kwh in 1970. The comparable figures for the local publicly owned systems, which charge lower rates, are 2,518 in 1950 and 9,015 in 1970. The higher usage by the public systems reflects the lower rates they charge, 1.00 cents per kwh in 1950 and 1.47 cents in 1970. *Ibid.*, p. 12.

The price of electricity charged by public agencies is lower because public power companies operate on a nonprofit basis and are exempt from federal income tax. In addition, their cost of capital is lower because municipal bonds are tax-exempt, they have preferential access to low cost federal hydroelectric power, and some of the larger public systems such as TVA and Bonneville are quite efficient.

149. *The 1970 National Power Survey.*

150. *Gas Facts*, A Statistical Record of the Gas Utility Industry in 1970, American Gas Association, Table 169.

151. This continued a long term trend in efficiency of coal use. In 1920, 3.1 pounds were needed and in 1935, 1.6 pounds to generate a kilowatt hour of electricity.

152. Lapp, "We're Running Out of Gas."

> At the consumer level, the cost breaks down like this: I burned 185,000 cubic feet of gas to heat my house along the Potomac during the last 12 months, at a total cost of $269. The well-head price of the gas was $37. The pipeline charges to bring it from Louisiana to Washington were about the same. There was $30 in sales tax and the Washington Gas Light Company charged me $165 for bringing the gas from its storage facilities to my home.

153. *Statistics of Income*, U.S. Treasury Dept., Internal Revenue Service, gives estimates of net income after taxes of $94 million in 1950 and $21 million in 1968. The IRS indicates, however, that their coverage of corporations engaged in bituminous coal mining is incomplete.

154. *Testimony of Robert Mitchell on Concentration by Competing Raw Fuel Industries in the Energy Market and Its Impact on Small Business*, Hearings Before the Subcommittee on Special Small Business Problems of the Select Committee on Small Business, 92nd Congress, p. 41.

> For instance, Pittston Company's return on sales (net income as percent of sales) was 6.9% in 1970, or substantially higher than its return of 4.1% in 1969. Similarly, West-

moreland Coal's return was 5.2% in 1970 as compared with only 1.5% in 1969. In addition, North American Coal's return on sales was 3.4% in 1970 as compared with 2.9% in 1969.

155. Keith C. Brown, ed., *Regulation of the Natural Gas Producing Industry*. Baltimore, Md.: Resources for the Future, Johns Hopkins University Press, 1972.

156. According to the Independent Petroleum Producers Association, independent oil producers have declined from about 10,000 in 1960 to 7,500 in 1970.

157. The 24 petroleum companies appearing among the 200 largest manufacturing companies of 1968 acquired assets totalling $7.9 billion during 1961-68, of which $2.5 billion or 31 percent were the merging of major companies with each other. Almost all of this $2.5 billion is accounted for by four massive mergers—Union Oil Company of California with Pure Oil; Sun Oil Company with Sunray DX; Atlantic with Richfield; and Phillips partial acquisition of Tidewater (marketing facilities in the Western states). In addition to mergers between major companies, petroleum companies also acquired $377 million of the assets of independent petroleum marketers and refiners.

158. See *Testimony of Robert Mitchell.*

159. By 1967 five major petroleum companies (Arco, Conoco, Gulf, Exxon and Sinclair, which merged with Arco in 1969) had combined holdings of 102 billion tons of recoverable coal reserves out of estimated 1,600 billion tons in the ground, and 2,491,000 acres of coal lands. It is of interest to note that in 1960 only one of the companies had any coal holdings, and these amounted to only 7.8 million tons of recoverable coal reserves and 4,524 acres of coal lands.

"Artificial Restraints on Basic Energy Sources," Memorandum prepared by Special Council for American Public Power Association and National Rural Electric Cooperative Association, April 26, 1971, pp. 64-65.

160. According to the House of Representatives Select Committee on Small Business 1971 Report on Concentration, the current position is as follows:

> Presently, the major oil companies account for approximately 84 percent of U.S. refining capacity; about 72 percent of the natural gas production and reserve ownership; 30 percent of the domestic coal reserves and over 20 percent of the uranium reserves and 25 percent of the uranium milling capacity. Further, the major oil companies are

acquiring oil shale and tar sands as well as water rights in many areas of the country.

CHAPTER 8

161. Private conversation with the author.

162. Garfield and Lovejoy, *Public Utility Economics*, p. 30.

163. TVA Act, Sections 10 and 11.

164. Bonneville Power Act, Section 837 (a).

165. Flood Control Act of 1944, Section 5.

166. *Public Utility Holding Company Act of 1935*, United States Securities and Exchange Commission, Public, No. 333, 74th Congress, Section 30, Washington, D.C.: U.S. Government Printing Office, 1964, p. 28.

> . . . to make studies and investigations of public-utility companies, the territories served or which can be served by public-utility companies, and the manner in which the same are or can be served, to determine the sizes, types, and locations of public-utility companies which do or can operate most economically and efficiently in the public interest, in the interest of investors and consumers, and in furtherance of a wider and more economical use of gas and electric energy; upon the basis of such investigations and studies the Commission shall make public from time to time its recommendations as to the type and size of geographically and economically integrated public-utility systems which, having regard for the nature and character of the locality served, can best promote and harmonize the interests of the public, the investor, and the consumer.

167. Phillips Petroleum Co. v. Wisconsin, 347 US 672 June 7 (1954).

168. Natural Gas Act, Section 5a.

169. In the 1961-64 period the FPC ordered $644 million in refunds and $156 million/year in rate reductions by natural gas producers and pipelines.

170. See *Summary of Congressional Hearings and Bills Proposed and Introduced in National Bureau of Economic Research*, Report of the Committee on Prices in the Bituminous Coal Industry, New York, 1938, pp. 124-26.

171. Atomic Energy Act. P.L. 555 79th Congress.

CHAPTER 9

172. Study by Representative Les Aspin (D-Wis.) released January 1, 1974. For an analysis of contributions to previous campaigns see Herbert E. Alexander, "Financing the 1968 Election." Lexington, Mass.: D.C. Heath and Company, 1971. Appendix G gives the occupational classification of contributors of $10,000 or more to the 1968 election. Oil accounted for almost 7 percent of the total, a much higher proportion than any other single occupation (except Investment Banking).

173. Rowland Evans and Robert Novak, *Lyndon B. Johnson: The Exercise of Power*. New York: The New American Library, 1966.

174. Philip H. Dougherty, "The Selling of the Energy Crisis." *The New York Times*, April 1, 1973.

CHAPTER 10

175. Of course some "waste" and inefficiency is unavoidable. The Second Law of Thermodynamics tells us that there are certain inherent inefficiencies in the conversion of matter from one form to another. Thus when we speak of reducing waste we mean increasing the percentage of our energy potential put to useful work by extracting more of the fuel from the ground and then using it more efficiently in heating our homes and running our cars and factories.

176. *Energy Research Needs*, Resources for the Future Inc., in cooperation with M.I.T.Environmental Laboratory, October, 1971., pp. VIII-38-39.

177. All glass municipal building in Tempe, Arizona, designed by architects Michael S. Kemper Goodwin.

178. Private communication from National Bureau of Standards.

179. The largest companies already have such teams at work. U.S. Steel, DuPont, and Eastman Kodak have launched aggressive energy conservation programs. DuPont is even selling its program to other companies. *Wall Street Journal*, October 5, 1972, p. 1.

180. Edmund Faltermeyer, "Metals, The Warning Signals Are Up," *Fortune*, October, 1972, p. 109.

CHAPTER 11

181. If you assume that 30 billion barrels of additional oil could be recovered by secondary and tertiary recovery and 30 billion barrels from heavy oils, these resources could support production of 6 million barrels a day by 1985.

182. Strip-mining recovers perhaps twice as much as underground mining. And while coal production from strip-mining is now about half of total production, the great bulk of our coal resource, some 70 percent, is in seams far enough below the surface that underground mining methods are likely to be required.

183. *Natural Gas Survey, Report of Gas Supply Task Force*, Federal Power Commission, 1973. Depending on the size of the reservoir, rate of production, type of fracturing and return on investment, the average wellhead cost for natural gas produced from the three potential Rocky Mountain fields is about 75 cents per thousand cubic feet.

See also: "An Analysis of Gas Stimulation Using Nuclear Explosives," B. Rubin, L. Schwartz, and O. Montan, May 1972.

184. *Natural Gas Survey, op. cit.*

185. *Ibid.*

186. "Abatement of Sulfur Dioxide Emission from Stationary Combustion Sources," Ad Hoc Panel on Control of Sulfur Dioxide From Stationary Combustion Sources. National Research Council, 1970, p. 53.

187. A successful pilot plant was completed in 1956 but shortly thereafter the federal research and development effort was abruptly halted by the Eisenhower Administration, which, with the strong backing of the oil industries, took the view that further development should be left to private investors.

188. *Wahakie Basin Corehole No. 1*, U.S. Bureau of Mines, open-file report, 1969.

189. *United States Energy, A Summary Review, op. cit.*, p. III-35.

190. *Ibid.*

191. "Energy Potential From Organic Wastes," U.S. Department of Interior, Bureau of Mines Information Circular 8549, 1972, p. 8.

192. *Ibid.*, p. 13.

193. U.S. Environmental Protection Agency data.

194. "Energy Potential From Organic Wastes," *op. cit.*, p. 13.

195. *Ibid.*

196. Pollution control is still a concern when the wastes are burned, but it could be a quite satisfactory way of disposing of solid wastes, especially when the present alternatives are to dump them in land-fills or burn them without any beneficial use.

197. *Forecast of Growth of Nuclear Power*, U.S. Atomic Energy Commission, March 1973.

See also *The National Power Survey*, 1972, Federal Power Commission, and *U.S. Energy Outlook*, National Petroleum Council, December 1972.

198. Another major inefficiency of current nuclear plants is the large amount of electric power required to enrich the uranium for use in nuclear reactors. For every 100 kilowatt hours produced by a nuclear power plant, about 5 kilowatt hours are required to enrich the uranium. This introduces another inefficiency into a woefully inefficient system. The breeder reactor would effectively eliminate this inefficiency because it does not require enriched uranium as fuel. An alternative enrichment process, the gas centrifuge process now being developed by the U.S.'s AEC and other nations as well, is much less energy intensive.

199. "Assessment of Nuclear Energy Options," statement by Dr. Manson Benedict, Professor of Nuclear Engineering, Massachusetts Institute of Technology, before the House Science and Astronautics Subcommittee on Science, Research, and Development, May 9, 1972, p. 10.

200. "The Potential for Energy Conservation," Executive Office of the President, Office of Emergency Preparedness, October 1972, p. F-7. According to the study, btu requirements per kilowatt hour of electricity dropped from approximately 15,000 in 1949 to about 10,400 in 1969.

201. If the fuel cell is to be competitive with the local electric utility, consumers must have access to fuel for the cell from an independent source. The natural gas industry, which already owns a pipeline system to most consumers, is a natural supplier, and the same pipelines could be used to deliver hydrogen when natural gas and synthetic gas runs out.

CHAPTER 12

202. *United States Energy, A Summary Review*, U.S. Depart-

ment of the Interior, January 1972, p. 57.

203. Robert W. Rex, "Geothermal Energy—The Neglected Energy Option," *Bulletin of the Atomic Scientists*, October 1971, p. 54.

David Fenner and Joseph Klarmann, "Power From the Earth," *Environment*, December 1971, p. 32.

204. *U.S. Energy Outlook*, New Energy Forms Task Group, Other Energy Resources Subcommittee, National Petroleum Council, 1972, p. 23.

205. Joseph Barnea, "Geothermal Power," *Scientific American*, January 1972, p. 70.

206. *Ibid.*, p. 77.

207. Fenner and Klarmann, "Power From the Earth," p. 33.

208. *U.S. Energy Outlook*, p. 24.

209. *United States Energy, A Summary Review*, p. 56.

210. John Banwell and Tsvi Meidav, *Geothermal Energy for the Future*, National Association of Atomic Scientists Meeting in Philadelphia, December, 1971.

211. John Noble Wilford, "New Plan Is Outlined for Tapping Geothermal Energy," *The New York Times*, Wednesday, June 21, 1972.

212. *A Feasibility Study of a Plowshare Geothermal Power Plant*, American Oil Shale Corporation and U.S. Atomic Energy Commission, U.S. Department of Commerce, Virginia, April 1971.

213. *United States Energy, A Summary Review*, p. 57.

214. Barnea, "Geothermal Power," pp. 73-74.

215. Less than 5 percent of the solar energy reaching the earth could supply the entire world population of almost 4 billion with the same per-capita level of energy consumption the U.S. enjoys today even if the energy were used at only 5 percent efficiency.

Energy Research Needs, A Report to the National Science Foundation, by Resources for the Future, Inc., in cooperation with MIT Environment Laboratory, October 1971, VIII-4.

216. Farrington Daniels, *Direct Use of the Sun's Energy*. New Haven, Conn.: Yale University Press, 1964.

U.S. Energy Outlook, p. 50

Small solar cookers (parabolic or cylindrical mirror concentrators) are sold commercially in India and elsewhere, and a large number of fairly large solar heaters and solar stills (water desalters) have been built in Japan, Africa, Australia, Israel and many of the Mediterranean nations. High temperature solar furnaces have been built, and there are several experimental solar-heated houses in existence.

217. *Energy Research Needs*, pp. VIII-12-13.

218. *U.S. Energy Outlook*, pp. XXXVI-38, 40-42.

219. *Ibid.*, p. XXXVI.
Energy Research Needs, pp. VIII-12-13.

220. *U.S. Energy Outlook*, p. 39.

221. In the Pacific Northwest electricity is very low in price because of the economical hydroelectric plants financed by the federal government. The price for electricity in that region is substantially lower than the cost of new capacity from steam plants required now that the low-cost hydropower has already been developed. F. A. Tybout and G. O. G. Lof, "Solar House Heating," *Natural Resources Journal*, Vol. 10, No. 2, April 1970, pp. 268-326.

222. Wilson Clark, "How To Harness Sunpower and Avoid Pollution," *Smithsonian*, November 1971, p. 16.

223. *Solar Energy Utilization: A Plan for Action*, Solar Energy Panel to Office of Science and Technology, Executive Office of the President, p. 18.

224. Lloyd G. Dehrer, "Solar Energy Heats Swimming Pool," *Heating, Piping & Air Conditioning*, October 1959, pp. 94-96.

225. *Solar Energy Utilization: A Plan for Action*, p. 13.

226. The local utility could play a major role in boosting solar heating and cooling if it would agree to finance the extra investment for the solar facilities and then recover its costs as part of the utility bill for supplemental heat or electricity, as the case may be. The utility might find this idea attractive because the solar equipment would then be part of the utility's "rate base"—the investment on which it is allowed to earn a return.

227. Unpublished Report of Conference of Solar Energy Experts held at the University of Delaware, October 8 and 9, 1971, p. 22.

228. *Ibid.*, p. 26.

229. The electricity from the solar cell would not make the house completely independent of the electric power company, however, unless an economical battery were developed to store electricity for use at night and on cloudy days. *Ibid.*, pp. 26-27.

230. For example, only 1 percent of the Sahara would be required to supply all of the electric energy in the entire world even with only a 5 percent conversion efficiency.

231. Gordon R. Ball, *Electric Energy Consumption and the Nuclear Opportunity*. Toronto: Canavest House Limited, 1972, p. 34.

232. Except for very short periods of up to 72 minutes twice a year near the equinoxes when the solar power plant would pass through the earth's shadow. This would occur in spring and fall when demand for electricity is not at its peak and would not appear to present a serious problem.

233. The power supply from the sun is a constant amount of some 430 btu/ft^2 hr on a surface normal to the sun's rays outside the earth's atmosphere, as the satellite would be positioned. At the earth's surface the absorption by the atmosphere combined with the fact that the sun is usually not directly overhead results in an average of only about 60 btu/ft^2 hr or only 1/7 of the concentration of solar energy available in space.
Energy Research Needs, p. VIII-3.
U.S. Energy Outlook, p. 38.
Yellot, "Solar Energy: Its Use and Control," pp. 101-110.

234. An official study of the technical feasibility of the satellite solar-power station is being made by NASA and should be completed in 1974.

235. *Solar Energy Utilization: A Plan for Action*, p. 48.

236. *U.S. Energy Outlook*, p. 52.

237. Hearings Before the Subcommittee on Research and Development of the Joint Committee on Atomic Energy, 86th Congress, 2nd Session, on Frontiers on Atomic Energy Research, March 22, 23, 24, and 25, 1960, pp. 346-358, especially page 346 where the Committee claimed jurisdiction because the sun is a thermonuclear reaction, and p. 358 where support for solar energy research was expressed by Mr. Hosmer, a member of the committee.

238. Dr. Hans A. Bethe published his theoretical studies in 1939, which revealed that the energy radiated by the sun and the

stars is produced by the fusion of hydrogen into helium. He was awarded a Nobel Prize for this work in 1968.

239. Statement by Melvin B. Gottlieb, Director of the Atomic Energy Commission's Plasma Physics Laboratory, *Business Week*, September 12, 1970, p. 80.

240. L. A. Artsimovich, "Controlled Nuclear Fusion: Energy for the Distant Future," *Bulletin of the Atomic Scientists*, June 1970, p. 48.

241. Tom Alexander, "The Hot New Promise of Thermonuclear Power," *Fortune*, June 1970, p. 97.
"Russia's H-bomb Tamers," *Business Week*, September 20, 1969, p. 7.

242. Moshe J. Lubin and Arthur P. Fraas, "Fusion by Laser," *Scientific American*, June 1971, p. 21.

243. Lowell Wood and John Nuckolls, "Fusion Power," *Environment*, May 1972.

244. Interestingly the AEC's laser research on fusion is classified, although the existence of the program and its level of funding are public information. The secrecy, presumably, is due to the potential military applications of the laser work as a trigger for H-bombs or otherwise. This, of course, raises a safe-guard question. Would a laser-powered fusion power plant provide its owner with the technology to make an H-bomb?

245. KMS Industries of Ann Arbor, Michigan, a research company, is using smaller lasers and concentrating on the size and makeup of the pellets. The government is challenging the company's patent applications on the ground that they are derived from the government's own work.

246. But it took only 3 years and $40 million to demonstrate the scientific feasibility of fission after the initial effort got underway in 1939.

247. David J. Rose, "Controlled Nuclear Fusion: Status and Outlook," *Science*, Volume 172, May 21, 1971, pp. 802-803.

248. Solving our waste disposal problem is another possibility suggested by fusion advocates. A fusion torch would theoretically be able to subject wastes to such intense heat that they would vaporize and then be reduced back to their basic elements and recycled. However, there is a good deal of skepticism as to the feasibility of this approach.

249. "The Hot New Promise of Thermonuclear Power," p. 132.

250. "Fusion Power: Progress and Problems," *Science*, pp. 802-803.

251. William C. Gough and Bernard J. Eastlund, "The Prospects of Fusion Power," *Scientific American*, February 1971, pp. 50, 60-63.

252. Wood and Nuckolls, "Fusion Power," p. 33.

253. There are ideas for direct conversion of fusion power to electricity but they require technological breakthroughs over and above those required to develop a fusion reactor.

254. Of course, the first plant will be built only as the climax of a long, hard research and development effort costing billions of dollars that will require extensive government support.

255. "The Hot New Promise of Thermonuclear Power," p. 136.

CHAPTER 13

256. Such a mandate could be enacted through amendments to the Federal Power Act and Natural Gas Acts that authorize the Federal Power Commission to regulate wholesale rates for natural gas and electric power and delegate retail rate fixing to the states. Congress has ample authority to provide guidelines for the states for such retail ratemaking.

257. Statement by Monte Canfield, Jr., before Senate Subcommittee on Anti-Trust and Monopoly, June 28, 1973.

CHAPTER 14

258. Kakuei Tanaka, *Building a New Japan*. Tokyo: Simul Press, 1970, p. 66.

259. *Ibid.*

260. Daniel Bell, *The Coming of the Post-Industrial Society*. New York: Basic Books, Inc., 1973.

261. John Maynard Keynes, *Essays in Persuasion*. New York: Harcourt, Brace and Company, 1933, pp. V-111.

> The day is not far off when the Economic Problem will take the back seat where it belongs and that the arena of the heart and head will be occupied, or reoccupied, by our real problems—the problems of life and human relations, of creation and behavior and religion.

262. "Global Constraints and New Vision for Development," a paper by Yoichi Kaya and Tutaka Suzuki, delivered at the Club of Rome Symposium held in Tokyo, October 1973.

263. Andrew Levison, "The Revolt of the Blue Collar Youth," *The Progressive*, October 1972, p. 38.

> We can gain a better understanding of how many factory workers feel if we imagine a job of typing a single paragraph over and over again from 9 A.M. to 5 P.M., five days a week. But instead of setting the pace yourself, your fingers must keep up with the typewriter carriage which moves at a steady rate from 9 A.M. 5 P.M. Your job is at stake if you can't keep up the pace typing that same paragraph over and over again. To go to the bathroom or use the telephone you must ask your supervisor's permission. You can be ordered to work overtime at the company's discretion or lose your job. If you are unlucky you work the night shift. All the time you face the grim conclusion that no matter how much you do the job there's no chance of promotion. Your job is a dead-end.

264. Bruce Boissat, "New Challenges Will Have to Be Found for Us," *The Evening Star and Daily News*, Washington, D.C., December 12, 1972.

Index